This original and exciting study offers a completely new perspective on the philosophy of mathematics. Most philosophers of mathematics try to show either that the sort of knowledge mathematicans have is similar to the sort of knowledge specialists in the empirical sciences have or that the kind of knowledge mathematicians have, although apparently about objects such as numbers, sets, and so on, isn't really about those sorts of things at all.

Jody Azzouni argues that mathematical knowledge really is a special kind of knowledge that must be gathered in its own unique way. He analyzes the linguistic pitfalls and misperceptions philosophers in this field are often prone to, and explores the misapplications of epistemic principles from the empirical sciences to the exact sciences. What emerges is a picture of mathematics sensitive to both mathematical practice and to the ontological and epistemological issues that concern philosophers.

The book will be of special interest to philosophers of science, mathematics, logic, and language. It should also interest mathematicians themselves.

"This is an account that rings true, and which philosophers will have to come to grips with because of the sure-footed basis that it has in its description of mathematical theory and practice."
– Arnold Koslow

T0245149

Metaphysical Myths, Mathematical Practice

Metaphysical Myths,
Mathematical Practice
The Ontology and Epistemology
of the Exact Sciences

JODY AZZOUNI
Tufts University

CAMBRIDGE
UNIVERSITY PRESS

CAMBRIDGE UNIVERSITY PRESS
Cambridge, New York, Melbourne, Madrid, Cape Town, Singapore, São Paulo

Cambridge University Press
The Edinburgh Building, Cambridge CB2 8RU, UK

Published in the United States of America by Cambridge University Press, New York

www.cambridge.org
Information on this title: www.cambridge.org/9780521442237

First published 1994
This digitally printed version 2008

A catalogue record for this publication is available from the British Library

Library of Congress Cataloguing in Publication data
Azzouni, Jody.
Metaphysical myths, mathematical practice : the ontology and epistemology of the
exact sciences / Jody Azzouni.
p. cm.
ISBN 0-521-44223-0
1. Mathematics–Philosophy. I. Title.
QA8.4.A98   1994
510´.1 – dc20   93-17660

ISBN 978-0-521-44223-7 hardback
ISBN 978-0-521-06219-0 paperback

For those who are gone:

N. A.   J. V.   E. V.

O. A.   B. D.   D. A.

E. R.   M. F.   J. B.

Contents

PART III
THE GEOGRAPHY OF THE A PRIORI

Acknowledgments

In one form or another, this book took five years. Consequently my debts are numerous and substantial: my thanks to Helen Cartwright, David Isles, Jerrold J. Katz, Juliette Kennedy, Arnold Koslow, Maureen Linker, Aaron Lipeles, David Pitt, Mark Richard, David Rosenthal, George Smith, and Stephen White. My thanks also to my philosophy of mathematics class, Fall 1992: they helped me to turn a final version of the book into a penultimate version.

I also want to thank the City University of New York Graduate Center for inviting me to be a visiting scholar for the academic year 1989–90. During that time I was able to write several articles that became part of the basis of this book. While at the Graduate Center, I was partially supported by a Mellon fellowship from Tufts University, for which I am grateful. Lastly, but not leastly, I want to thank the philosophy department at Tufts University for providing a near-perfect environment in which to *do* philosophy.

PART I

Mathematical Practice and Its Puzzles

It seems clear that what is needed in Philosophy of Mathematics is work that is *philosophical* and not primarily technical.

Hilary Putnam

§1 Metaphysical Inertness

Here is a portrait of mathematical practice: The mathematician proves truths.

In bygone days, such truths were couched in the vernacular augmented with a small list of technical terms. The proofs were (more or less) detailed arguments; these arguments were (more or less) valid, where validity was understood informally according to the standards of the time; and these arguments were surveyable, provided one had the training.

How different things are these days is a bit hard to determine. Most proofs are still detailed (and surveyable) arguments couched in the vernacular augmented with a (somewhat larger) list of technical terms. Some proofs, however, are the results of computer calculations and are not surveyable.[1] First-order canons of validity have been formalized, but this has been achieved by constructing first-order formal languages that (theoretically speaking) seem able to replace natural languages as the medium for mathematics.[2]

However, there is some debate about the faithfulness of the first-order mirror of mathematics. One problem is that many mathematical notions are simply not first-order definable, 'finite' and 'infinite' being obvious examples. Another (related) problem is the Löwenheim–Skolem theorem: Significant first-order theories admit of unintended nonisomorphic models.

For these and other reasons, some philosophers and logicians think that the line between logic and mathematics should not be drawn at the boundary of first-order logic. They believe that the first-order canonization of inferential principles does not exhaust inferential principles employed in mathematical practice (and, I might add, in ordinary life). Some even question the sense of such a boundary, and in doing so, may question the topic neutrality of "logic." Consequently, there is a debate about which such formal languages, if any, are the most suitable for mathematics, and there is even some debate about their relevance to mathematical practice.[3]

---

1 The epistemological implications of this are debatable. See, for example, Tymoczko 1979, Detlefsen and Luker 1980, and Resnik 1989.

2 The first-order approach alluded to here takes first-order logic as topic neutral and then studies different branches of mathematics by axiomatizing their notions in a first-order language. One may distinguish a further foundationalist approach that interprets mathematics in a first-order set theory. Classical mathematics can be handled in first-order ZFC; but this is not true of all contemporary mathematics. For example, Feferman (1982, p. 238) notes that the informal general theory of *mathematical structures* cannot be captured in ZFC. Quine is the classical practitioner of the first-order set-theoretic approach. See Quine 1940, 1969a.

3 The literature here is enormous. A taste: Barwise and Feferman 1985, Benacerraf 1985, Boolos 1975, 1984, Myhill 1951, Quine 1970, Shapiro 1985, 1990, Wagner 1987.

In any case, such formal languages are (practically speaking) too awkward for the ordinary mathematician to use in research. Modern mathematics, for better or worse, is still practiced in natural languages.

Although the primary focus of the mathematician seems to be the mathematical *truth,* the language of mathematics contains noun phrases and predicate phrases just like the language of any other subject matter; and what mathematical truths are about, naively, seems to be whatever these phrases refer to: numbers, relations, functions, sets, Banach spaces, and so on, through the range of objects mathematicians study.

Let us call *linguistic realism* the doctrine that takes mathematical talk at face value. Linguistic realists see the semantics of mathematical discourse as indistinguishable from discourse in other (empirical) contexts. In particular, linguistic realists take the semantics of mathematical discourse to be pretty much the same as the semantics for the rest of one's language: They define the truth of mathematical statements in the same way they define the truth of empirical statements, and they treat the mathematician's tendency to speak about the objects he or she is studying in the same way they treat the empirical scientist's tendency to speak about the objects *he* or *she* is studying.[4]

What is striking, however, is that the practice of mathematics, when compared to the practice of (one or another) empirical science, *seems* to support a picture of mathematical objects that makes them radically different from other (empirical) objects. *Except for* whatever (inexplicable) process the mathematician engages in to gain access to such objects, they don't seem to interact with us, or with the rest of our world. Their apparent metaphysical inertness, except as a semantic basis for mathematical truth, helps motivate claims that such objects are acausal, abstract and exist outside of space and time, and perhaps contributes to common intuitions about the *necessity* of mathematical truths.[5]

It is worth pointing out that the impression that mathematical objects, if there are any, are metaphysically inert is not due to what has been called the *aprioricity* of mathematical knowledge: the claim on the part of some philosophers that access to mathematical truth is independent of experience. Rather, what is at issue is the less tendentious observation of the difference between epistemic practices in the

---

4 Linguistic realism is certainly an important component of traditional platonism, and indeed, Benacerraf (1973) calls linguistic realism a " 'platonistic' account" (p. 410). But there are other elements in traditional platonism that go beyond linguistic realism, which is why I shy away from calling the doctrine described here *platonism.* See Part III, Section 6, where I discuss this further.

5 See, for example, Russell 1912, especially chapters 8, 9, and 10. Also see Hardy 1967.

empirical sciences and in mathematics. A crucial part of the practice of empirical science is constructing means of access to (many of) the objects that constitute the subject matter of that science. Certainly this is true of theoretical objects such as subatomic particles, black holes, genes, and so on. However analogous theoretical work in higher set theory may seem to be to theoretical work in the empirical sciences,[6] this disanalogy remains. Empirical scientists attempt to interact with most of the theoretical objects they deal with, and it is almost never a trivial matter to do so. Scientific theory and engineering know-how are invariably engaged in such attempts, which are often ambitious and expensive. Nothing like this seems to be involved in mathematics.[7] *At best* the process the mathematician uses to engage mathematical objects seems to be like introspection. This may explain the nineteenth-century temptation to reduce mathematical objects to psychological ones.

The metaphysical inertness of mathematical objects gives rise to a twin pair of puzzles that have been the implicit focus of most contemporary work in the philosophy of mathematics: Given this inertness, how do we refer to such objects, and how do we know the things we know about such objects? These puzzles have profoundly influenced the direction of contemporary philosophy of mathematics primarily by driving philosophers away from linguistic realism.

One aim of this book is to defend a version of linguistic realism. Let me give a sketch of how I propose to do this. One may argue that the presence of noun phrases and predicate phrases in the language mathematicians use is only prima facie evidence that mathematicians are committed to mathematical objects. This evidence is primarily linguistic: It is that the same locutions occur in mathematical discourse as occur in other discourses; and in other discourses such locutions are taken to be ontologically committing. So, for example, just as we say, "there are chairs and tables," we also say, "there are even and odd numbers." And just as we say, "there are more functions from the natural numbers onto the natural numbers than there are polynomial

6 Quine, early and late, has stressed the resemblance between high-level theory in the empirical sciences and high-level theory in mathematics. It is the forging of ontological commitments that goes on in both sorts of scientific theorizing that the analogy is partially rooted in. See, for example, his 1954, p. 122, his 1951a, and his 1981a. Quine, perhaps, was the first to empiricize mathematics this way. Mill's assimilation of mathematics to the empirical sciences does not take account of the higher reaches of theory in either case, near as I can tell. The tendency to assimilate mathematics to the empirical sciences this way has been taken up by many philosophers, in some cases independently of Quine. See, for example, Putnam 1975a, Lakatos 1976a, Kitcher 1984, and Maddy 1990. Parsons (1979–80) demurs to some extent, for reasons I take seriously (and discuss eventually), but he does not mention this particular disanalogy.

7 Of course empirical scientists talk about mathematical objects too. But their epistemic attitude toward them seems identical to the one mathematicians take.

expressions," we say, "there are more grains of sand on Jones Beach than any sensible human being can count."

But prima facie evidence is, well, *prima facie*. Even if one cannot avoid certain ontological commitments by regimenting in a first-order formal language and rephrasing (as we can with talk of, e.g., 'sakes'), still if the apparent commitments generate enough difficulties, one may attempt more radical ways of avoiding them. One may introduce primitive modal notions, substitutional quantification, or other things.[8]

My point here is simply to stress where the burden of proof lies. From the point of view of linguistics, the use of noun phrases and predicates in mathematical language is *indistinguishable* from the use of noun phrases and predicates in other contexts. Thus, it is only the *philosophical* puzzles that I mentioned earlier that motivate nominalism, modalism, and the various other strategies recently employed by philosophers of mathematics to avoid linguistic realism.

I don't want to give the impression that I think anyone is under illusions on this point. Philosophers promoting nominalist-style programs invariably start with at least a cursory nod toward classical discussions of the problems of linguistic realism (as found, say, in Benacerraf 1973). But this shows something that philosophers of mathematics offering alternatives to the realist position rarely mention: *Linguistic realism is a position that we must be driven away from by purely philosophical considerations.*

This gives the defender of linguistic realism an obvious strategy: He or she need only undercut the philosophical puzzles that make philosophers uneasy about the position. It is an obvious strategy, although not an easy one. The crucial move will be to break the stranglehold that the twin set of puzzles has had on contemporary philosophy of mathematics. In the process of doing this, I will show that in the shadows of these puzzles lurk other closely related puzzles which are largely overlooked in the literature. We will see that these other puzzles point the philosopher of mathematics who takes them seriously in a very different direction from the one that most contemporary philosophers of mathematics have traveled.

§2   Metaphysical Inertness and Reference

Much of the recent work in the philosophy of language has been concerned with a question that goes well beyond issues in the philosophy of mathematics proper: How do we refer to objects at all? The point of the question is straightforward. Reference does not take place by

8 See Putnam 1967, Gottlieb 1980, Hellman 1989, Field 1980, 1989.

magic; it is not something that just happens when we talk to each other.[9] We have to tell a story of how the terms we use refer to what they refer to, and this story is not supposed to be nonnatural. That is, whatever the story is, it must be consistent with our current scientific picture of the sort of creatures we are. We call the restraint vaguely described here the requirement of *naturalized semantics*.

A popular class of answers to the question of how we refer are "causal" theories of reference.[10] Although not without substantial problems, the picture they give is appealing because of their compatibility with the demands cited in the previous paragraph. We are creatures with sense organs, and these sense organs operate by responding to certain things by (some of) their causal effects. In one version of the causal theory of reference, the referential relation *just is* a causal relation of a certain sort. Clearly such a view, on pretty much any version, cannot be used to explain how we refer to mathematical objects if they are taken to be acausal. I call this (not without precedence) the *puzzle of referential access to mathematical objects.*

Strategies for solutions to this puzzle fall into two classes. First, we may claim there are referential resources the above description of the problem overlooked: I call such strategies *ontologically conservative.* Second, and more radically, we may claim either that the purported mathematical objects have not been described correctly (and when they are described correctly, they turn out to be referentially accessible), or that we can replace the traditional metaphysically inert mathematical object with something else that is referentially accessible. I call strategies of the second sort *ontologically radical.*

Examples of ontologically conservative strategies are ones that utilize higher-order logic, informal methods of reference, or intellectual intuition. Examples of ontologically radical strategies are ones that take mathematical objects to be located in space and time (and therefore accessible by the same referential methods we use to refer to anything else in space and time) and various types of modalism – the taking of modal idioms as primitives and mathematical reference as, strictly speaking, concerned with possibility rather than actuality. By way of making this typology exhaustive, nominalism, the denial of the existence of mathematical objects altogether, may be regarded as ontologically radical.

Positions in the philosophy of mathematics are rarely pure concoctions involving only strategies in one class or the other. For example,

9 I borrow this way of putting matters from Putnam 1981, pp. 1–5.
10 Here, too, the list is enormous and I can give only a taste: Kripke 1980 and Putnam 1975b, 1975c are classics. Further articles of interest may be found in Schwartz 1977. Also see Devitt 1981. A rather close sibling of the causal theory is the attempt to ground semantic content on *information.* See Dretske 1981 and Fodor 1987.

Shapiro's position (to be discussed in some detail shortly) uses both second-order logic, which is ontologically conservative, and the ontologically radical move of substituting *structures* for the traditional platonic objects as the subject matter of mathematics. Nevertheless, to the extent that the literature allows it, in this and the next section I focus primarily on ontologically conservative strategies. This is because, prima facie, they seem to be easy responses to the puzzles arising from metaphysical inertness. It is important, therefore, to see that they will not do, and why. Later on, I consider, although fairly briefly, certain ontologically radical strategies. I do not intend to give detailed refutations of positions utilizing radical strategies because, as I have mentioned, my general aim is to show that the puzzles that linguistic realism in mathematics seems to offer are not genuine; the burden-shifting considerations of Section 1 will then put linguistic realism at an advantage with respect to any position that relies on a radical strategy.

Turning, therefore, to conservative strategies, the first one it is natural to discuss is one often mentioned (although rarely endorsed): the claim that we have referential access to mathematical objects in some special way. For example, it might be suggested that we possess the capacity for a kind of intellectual intuition, analogous to sense perception perhaps, and that such powers act as a foundation for our capacity to refer to mathematical objects. Because the position is so rarely seriously argued for, it is hard to evaluate what precisely it comes to. In some versions mathematical objects have causal powers, although the causal powers needed are of a type that goes beyond what is naturalistically acceptable. Other versions depict intellectual intuition as working in some other way that enables us to grasp metaphysically inert objects without interacting with them.[11] In any case, most philosophers are prone to regard such strategies as not in the spirit of naturalized semantics unless the capacity for intellectual intuition can be shown to be compatible with the powers that creatures with brains and sense organs like ours have.

At this point it is natural to mention another conservative strategy: to argue that reference to mathematical objects is inherently an informal matter.[12] Sometimes Gödel's theorem is offered as a justifica-

---

11 A Kantian version of this move may be found in Katz 1981.
12 Myhill 1951 is a commonly cited example of this sort of view. It is often coupled with certain versions of the just mentioned position that access to mathematical objects is through (nonnatural) intuition. Gödel is sometimes taken to have believed this, although Tait (1986, pp. 364–5) convincingly suggests otherwise. But one finds an inclination toward this view even among philosophers who otherwise do not seem sympathetic to nonnatural intuition. See, for example, Pearce and Rantala 1982 and Tymoczko 1990.

tion for the claim that our grasp of mathematical objects transcends any possible formalization. But one can use the more elementary model-theoretic results mentioned earlier for the same purpose. And if one argues that mathematical practice inherently presupposes a grasp of mathematical objects that goes beyond what formal means seem to allow us, in effect what one has is a transcendental deduction of the capacity to refer to mathematical objects. To the extent that such an argument amounts to the claim that problems with reference should not be taken to license revision of mathematical practices, I am sympathetic. But it should not be taken as a genuine explanation of how we refer to mathematical objects.[13]

This brings us to the last conservative strategy I will discuss. We can try to pick out objects and types of objects by descriptions, that is, by lists of truths that hold uniquely of them. So, for example, we might consider a countable first-order language (predicates, first-order variables, first-order quantifiers, connectives, etc., the cardinality of the vocabulary at most $\aleph_0$), and a recursive set of truths (axioms) that we take to hold of, say, the natural numbers. Such descriptions pick out the desired objects, if they do, not by use of causal connections between the terminology and the objects referred to, but by the meaning we invest in the logical idioms (e.g., in the quantifiers and logical connectives).

The first-order version just described seems clearly inadequate: Because of the Löwenheim–Skolem theorem, any countable first-order set of sentences with a model of infinite cardinality has models the size of any infinite cardinal, regardless of the cardinality of the intended model.[14] Furthermore, even when the intended model is countable, unintended models of the same cardinality are possible. The most notorious example, of course, is first-order Peano arithmetic.

Many of these noncategoricity results vanish when one ascends to more powerful logics such as diminished second-order logic, ω-logic,

---

13 Shapiro (1985, p. 720) expresses similar sentiments. He draws the conclusion that the categoricity second-order languages bring is crucial to explaining how mathematicians communicate to each other: Utilizing only first-order languages, and without a preestablished harmony, we cannot be sure mathematicians apparently talking about the same objects are tuned in to models that are even isomorphic to each other. Eventually I show this second-order strategy will not work. See Section 3.

14 A precise statement of the theorem, its proof, and interesting strengthenings may be found in Chang and Keisler 1973. They call it the Löwenheim–Skolem–Tarski theorem for obvious reasons. I retain the shorter nomenclature simply because that is how the theorem is commonly referred to in the philosophical literature. Of course one or another version of the result can also be found in most logic textbooks, for example, Shoenfield 1967, Enderton 1972, and Boolos and Jeffrey 1989.

higher-order logics, or logics with quantifiers that are more cardinal-sensitive than the first-order quantifiers.[15] However, one must not rejoice too quickly at the apparent banishment of this problem. Any evaluation of an approach for successfully referring to mathematical objects by augmenting the logical idioms at our disposal must take account of the naturalized constraints mentioned in the first paragraph of this section. The idioms of *any* formalized language must be scrutinized to make sure they do not presuppose illegitimate (i.e., non-naturalized) access to mathematical objects.

For example, one could certainly add to the first-order predicate calculus a one-place logical predicate 'N' taken to hold of all and only the natural numbers, much as ' $=$ ' is standardly added to the first-order predicate calculus as a logical constant. Such a device clearly violates naturalistic constraints, as no indication is given for how it achieves this referential miracle.

Only slightly less transparent is the case of ω-logic. Here a different style of quantifier and variables, which are taken to range over the natural numbers, is added to first-order logic. ω-logic has interesting properties, but one of them cannot be that it supplies a solution to the question of how we refer to the natural numbers, because that ability is simply presupposed in the new logical idioms added to standard logic.[16]

Characterizing types of mathematical objects (that, informally speaking, we seem successfully to refer to) within one or another formal language has generated quite a bit of controversy. There is no consensus on whether the problem of categoricity is solved by ascending to formal systems with greater expressive powers than those possessed by the first-order predicate calculus. I argue in Section 3 that, subject to the naturalistic provisos mentioned earlier, it is unlikely that any method of augmenting the first-order predicate calculus with logical idioms of greater power solves the problem of reference to mathematical objects if the first-order predicate calculus cannot do it on its own.

I should add that categoricity, in any case, does not necessarily gain us referential access in the sense we might naively expect it. For isomorphic models, where the objects are permuted, always satisfy the same theorems. That is to say, augmenting the expressive power of the first-order predicate calculus in the ways described above cannot fix particular mathematical objects as referents for particular terms.

15 Definitions for these may be found in Shapiro 1985. See also Corcoran 1980, Barwise and Feferman 1985, and Van Benthem and Doets 1983.
16 Shapiro (1985, p. 733) writes that ω-languages "cannot be used to show, illustrate, or characterize how the natural number structure is itself understood, grasped, or communicated," since "they assume or presuppose the natural numbers."

For example, any abstract object – for that matter any object at all – can play the role of 0 in a model that satisfies second-order Peano arithmetic.

Structuralists, such as Shapiro, are willing to accept this sacrifice. Because isomorphism is the best we can get with second-order logic, one gives up this much of naive mathematical practice: When talking about *the* natural numbers, one can at best be talking about *structures,* something that models of second-order Peano arithmetic have in common.[17] This is ontologically radical and therefore forces Shapiro and other structualists of his ilk to rely on Benacerraf-style objections to linguistic realism.

But let us leave aside the structuralist aspect of these positions for the time being and focus on whether higher-order logics gain us legitimate referential power over and above what first-order logic seems to offer. This is the topic of the next section, which I have tendentiously titled:

§3 The Virtues of (Second-Order) Theft

Shapiro is not alone in believing that higher-order logics – for example, second-order logic – have certain virtues over and above the first-order predicate calculus.[18] For our purposes, the primary virtue at issue is that, in using it, categorical definitions of certain mathematical structures, such as the natural numbers, become available. This perfectly ordinary mathematical fact becomes philosophically significant when it is argued that the problem of referential access is solved by using second-order logic. Crucial to this position are the kinds of claims one finds in Shapiro 1985 – for example, that in utilizing second-order logic one is not surreptitiously presupposing a greater referential grip on the mathematical objects than one is

17 See Shapiro 1989 for details on his structuralism. Structuralism, of one sort or another, is fairly popular. Other philosophers who have adopted structuralist strategies are Resnik (1981, 1982) and Hellman (1989). Hellman combines his structuralism with second-order logic (and modalities). Resnik is suspicious of going second-order (see Resnik 1988). Also see Parsons 1990.
18 See Shapiro 1985, 1990, Field 1989, Hellman 1989, Hodes 1984, among others. It is also worth adding to the list those philosophers who have invoked higher-order logics in debates with Putnam (1977), where the specter of Löwenheim–Skolem is posed as a puzzle for theories of reference in nonmathematical contexts, for example, Hacking (1983).
  The second-order strategy is often combined with strategies for spiriting away mathematical commitments altogether – namely, modalism and fictionalism. I am not concerned with these moves here but only with the suggestion that ascent beyond first-order idioms eliminates the problem of referential access (modulo isomorphism).
  I should add that what is meant here is second-order logic with the standard interpretation, that is, where the second-order quantifiers range over *all* the subsets of the domain.

entitled to. Shapiro (1985), as we've seen, argues that ω-logic does presuppose our capacity to refer to the natural numbers, and he also claims that first-order set theory, if taken to be interpreted in the standard model, presupposes our capacity to refer to the standard model.

What I want to do here is show that, in fact, exactly the same is true of second-order logic, although this is neatly masked by the second-order terminology. Consequently, ascending to second-order logic cannot solve the puzzle of referential access, if it is not already solved by merely invoking the standard interpretation in the first-order case.

I should add that Shapiro (1985, 1990) is not merely concerned with the problem of referential access. He wants to show that mathematical practice is best understood in second-order terms rather than in first-order terms. That issue is not on the table now. My only claim is that no one should think that because the use of second-order logic enables us to rule out nonstandard models of arithmetic, this has *explained* how we do rule out these models – not if the mere invocation of the standard model (or models isomorphic to such) in the first-order context can't do the job by itself.

A few remarks about the first-order predicate calculus are in order. Normally, the connectives are entirely fixed in their interpretation, whereas the predicates, constants, and quantifiers are only partially so fixed. What is allowed to vary *arbitrarily* are the domains for the quantifiers, and within those domains, what the ranges of the predicates are taken to be, as well as what individuals in the domain the constants are mapped to.

The expressive weakness of the first-order predicate calculus is due to the variations admissible in interpreting quantifiers, predicates, and constants. A model for a set of sentences S is simply a domain in which the terminology just mentioned is given an interpretation (subject to the constraints above), which satisfies S. The problem, of course, is that in general S is so easily satisfied.

We can imagine, however, restricting the class of models in certain ways. Here is an explicit definition. Let S be a set of statements, and let A be the class of models of S.[19] A *truncation* of A is a class of models of S that does not include every model in A. A *truncated* model theory for the first-order predicate calculus is a class of models that does not include every model.

We can characterize truncated model theories in two ways. We can require every model in the truncation to contain something; for example, that every model in the truncation contain infinitely many ob-

---

19 This discussion is taking place against an implicit set-theoretic background (for it is from set theory that the models are drawn). In general, the collection of models that satisfy a theory is too large to be a set; so I use "class" here quite deliberately.

jects. Or, we can require one or more of the constant symbols or predicate symbols to hold of particular sorts of sets; for example, that the predicate symbol 'P' hold only of two-membered sets in any model of the truncated model theory.

We call a first-order logical system with a truncated model theory a *truncated first-order logic*, or a *truncated logic*, for short.

Now here is the crucial point. For an absolutely trivial reason, truncated first-order logics have greater expressive power than the first-order predicate calculus. In particular, given judiciously chosen truncations, we will find that every notion is "truncated first-order definable." Nevertheless, without considerable argument, there is no reason to assume that adopting a truncation is a legitimate method for solving the problem of referential access. That is to say, if I use a truncated model theory in which the natural numbers are "truncated first-order definable," this certainly gives us no reason to believe that we have access to models of arithmetic that are isomorphic to the standard model – for we have been given no explanation for how we are able to exclude the models not in the truncation.

There is an easy method for translating certain logics, ones built on the first-order predicate calculus with additional idioms not defined in terms of the first-order vocabulary, into truncated first-order logic. I'll illustrate it with a version of the second-order predicate calculus containing n-place predicate and function symbols, but where second-order quantification occurs only in one-place predicate places.[20]

---

20 By sticking to this case, I am avoiding certain details. If we have second-order quantifiers ranging over functions or n-place relations (as we do with the general case), then model-theoretic apparatus must be added to handle them ('S', 'E', A, and B forthcoming will not do the job). But this changes nothing essential to the points I want to make with this construction; for if my points hold of the case I consider, then they hold a fortiori of the stronger cases.

A few words about other logics. The method described can be easily extended to define truncated logics corresponding to full second-order logic as just mentioned. It can also be applied to higher-order logics of all orders, predicative logic, as well as to logics with generalized quantifiers (e.g., "there are uncountably many As"). I don't see, offhand, how to extend the method to provide translations for infinitary languages. But, of course, whatever additional expressive power such languages have can be captured by one truncated logic or another. Finally, it can handle diminished second-order logic also – for example consider Corcoran's (1980) saLK's. A truncated logic similar to the one I explicitly give can be used, provided we restrict the class of wff's so that the predicate 'S', and the first place of 'E' only admit of universal quantification, and where the quantifier must occur at the beginning of the wff.

Finally, I make no claims for the technical originality of truncated model theory. It is really nothing more than the translation of higher-order logic into first-order logic via the relativization of quantifiers. A straightforward presentation of Henkin's general models using this route may be found in Van Benthem and Doets 1983 (section 4.1). My adaptation focuses only on the pertinent transformations for standard interpretations because those interpretations are the relevant ones for our philosophical purposes.

First an observation about language. The second-order language I will consider is two-sorted with respect to quantifiers: $\forall x$, $\exists x$, $\forall X$, $\exists X$; it has individual constants: a, b, c, . . .; individual variables: x, y, z, . . .; predicate constants: P, Q, R, . . .; and (one-place) predicate variables: X, Y, Z, . . . The $(2m)$ truncated language corresponding to this language has one style of first-order quantifier: $\forall x$, $\exists x$;[21] it has the same individual variables, individual constants, and predicate constants as the second-order language it is being designed to mimic, but no predicate variables. In addition, it has two new "logical constants": S, E, the former one-place, the latter two-place, which will be given fixed interpretations across all models.

Here is how we define the truncated model theory, the logic of which I call $(2m)$*truncated first-order logic*. We admit only models with the following properties into the truncation: Any such model M has a domain that can be exhaustively and disjointly divided into two classes A and B, where A is arbitrary and B $= \mathscr{P}$A. Furthermore, we fix the references of the two logical constants 'S' and 'E', so that in every model 'S' is mapped to B, and 'E' is mapped to the set of ordered pairs (a, b) such that a $\in$ A, b $\in$ B, and a $\in$ b. Finally, in every admissible model, any individual constant is mapped into A and the extension of every n-place predicate is a subset of A^n.

$(2m)$truncated model theory is meant to be equivalent (in a sense we will define below) to our version of monadic second-order logic with standard semantics. If one wants to capture a nonstandard interpretation for second-order logic, such as Henkin semantics, one needs (roughly) to substitute as the possible domain of B not the entire power set of A but possible subsets thereof. I forgo details.[22]

Here are some examples of how the translation of sentences of our second-order logic into the language of $(2m)$truncated logic must go. Consider the second-order induction schema for PA.

$$\forall X \ ([X0 \ \& \ \forall x(Xx \rightarrow Xsx)] \rightarrow \forall xXx).$$

This translates as

$$\forall x(Sx \rightarrow ([E0x \ \& \ \forall y(Eyx \rightarrow Esyx)] \rightarrow \forall yEyx)).$$

Consider next a second-order formulation of the statement that a relation R is *well founded:*

$$\forall X[\exists x(Xx \rightarrow \exists x(Xx \ \& \ \forall y(Xy \rightarrow \neg Ryx)))].$$

---

21 Or more precisely, "$(2m)$ truncated first-order language" – '$(2m)$' because the logic this is a language of is equivalent in power to the (monadic) second-order logic studied above, and 'first-order' because that is (syntactically) what it is. I trust the terminology and the abbreviations I help myself to will not mislead.

22 For those who want details on Henkin semantics, see Van Benthem and Doets 1983, Henkin 1950, Shapiro 1990, or Boolos 1975, among others.

This translates as

$$\forall z(Sz \rightarrow [\exists x(Exz \rightarrow \exists x(Exz \,\&\, \forall y(Eyz \rightarrow \neg Ryx)))]).$$

The translation method is easy: We treat this version of second-order logic as a two-sorted first-order logic and eliminate the additional quantifiers by relativizing to predicates. The truncated model theory must not be forgotten, however, for in no way can the standard first-order predicate calculus capture the expressive power of this second-order logic.

Just as a one-one translation of this second-order logic into $(2m)$truncated logic is available, so too is a one-one mapping of this second-order model theory onto $(2m)$truncated model theory. Here is the definition. A model A_s of second-order model theory is mapped to that model A_t of second-order truncated model theory in such a way that

(1) If d_s is the domain of A_s, then $d_s \cup \mathcal{P}d_s = d_t$.
(2) 'S' is mapped to $\mathcal{P}d_s$
(3) 'E' is mapped to $\{(a, b) \mid a \in b \,\&\, a \in d_s \,\&\, b \in \mathcal{P}d_s\}$.
(4) Individual constants are mapped to the same items in d_t as they are mapped to in d_s. n-place predicates are mapped to the same subset of d_t as they are in d_s.

The interesting metalogical properties, among which may be found versions of completeness, compactness, and Löwenheim–Skolem, arise from the interplay of the model theory and the language of the logic under study. It is easy to see that this second-order logic and $(2m)$truncated logic have identical properties in this respect. In particular, compactness, Löwenheim–Skolem, and completeness all fail, and the two logics have identical capacities to characterize infinite structures.

What about second-order logic where quantification takes place over n-place predicates and functions? There are two ways to go. First, we introduce subdomains $B_1, B_2, B_3, \ldots, B_n, \ldots$, which contain, respectively, all one-place relations, two-place relations, three-place relations, and so on, defined on A. And we introduce one-place predicate constants $R_1, R_2, R_3, \ldots, R_n, \ldots$, and a sequence of predicate constants: $E_1, E_2, \ldots, E_n, \ldots$, where each E_i is an $i+1$-place predicate. The models require that each R_i be mapped to B_i (so R_i is mapped to the set of all i-place relations on A) and that each $E_i a_1, \ldots, a_{i+1}$ has the fixed interpretation that the sequence of items $(a_2, \ldots, a_{i+1})$ is a member of a_1. Something similar may be done with functions.

If this seems too busy, an alternative is just to modify the model theory so that the logical constant R is mapped to the superstructure on A, take E to be ϵ, and then ape the definitions of relations and functions as is normally done in standard set theory. I should add that I certainly am *not* claiming that truncated first-order logic is easier to work with than second-order logic – on the contrary.

A moral or two is in order. First, it has been said that second-order logic (with standard semantics) is "set theory in sheep's clothing" (this phrase was coined, I believe, by Quine[23]). Now this remark, as has been noted by other philosophers, must be strongly qualified; for interpreted in the most direct way, it is simply *not* true. Second-order logic with standard semantics is hardly set theory, if by that phrase is meant *first-order set theory*. They simply don't have the same models. First-order set theory is open to multiple nonstandard models in which 'ϵ' is not mapped to the membership relation. This is not true of 'E'. Nor is it quite fair to identify second-order logic with *interpreted* first-order set theory, where the model fixed is the standard model, for even full second-order logic is too weak by comparison. Rather, second-order logic with standard semantics is *equivalent* in the strongest sense of the word to the truncated first-order logic indicated above, for these *do* have the same models, metalogically considered. But notice that it is not quite accurate to regard E as the membership relation over $A \cup \mathscr{P}A$ in $(2m)$truncated logic. This is because E is insensitive to whatever set-theoretic relationships there are that hold among the elements of A. Analogous remarks apply to the notation required in truncated logics that capture full second-order logic. Nevertheless, E captures enough of set theory for set-theoretic issues invariably to intrude in the analysis of the metalogical properties of second-order logic.

Next, the equivalences I have shown to hold between various augmented logics and truncated first-order logic make it clear that there is no justification for the claim that such augmented logics solve the accessibility puzzle; for example, it is obvious that it is a simple act of fiat to fix the language and the model theory in the way I have to get $(2m)$truncated logic. But I need to explain why the *impression* of fiat so obvious in this case is absent when most philosophers consider second-order logic with standard semantics.

Part of the reason is that when second-order logic is discussed, the implicit alternative to the standard semantics is the Henkin interpretation. The second-order quantifiers in that interpretation of the monadic case, for example, do not range over the *full* power set of the domain. Naturally it becomes possible to argue that it is perfectly ac-

---

23 See Quine 1970, p. 66. Discussion of this issue may be found in Van Bentham and Doets 1983, Boolos 1975, and Shapiro 1985.

ceptable (and understandable) to take second-order "all" to mean "ALL," and the opponent of second-order logic finds herself in the position of attempting to say that she doesn't understand "ALL."[24]

The debate remains deadlocked at this level because the acts of fiat are neatly tucked away in the second-order quantifier and the grammatical relation of predicate constant and variable to individual constant and variable. Let me specifically illustrate these in the monadic case I have given.

Normally, first-order logic is not taken to commit one to the extensions of the predicates as entities over and above what such extensions are composed of, but only to what the quantifiers range over. Thus the notation 'Pa', which is a one-place predicate symbol syntactically concatenated with a constant symbol, is not taken to contain an (implicit) representation of ϵ.

However, as soon as we allow ourselves to quantify (standardly) into the predicate position, this is precisely how syntactic concatenation *must be* understood.[25] Furthermore, and this is striking (or ought to be), syntactic concatenation in these contexts is *not* open to reinterpretation across models – it is an (implicit) logical constant.

This explains, I think, both why, when we make the concatenation involved here explicit in terms of 'E', it is clear that 'E' must be fixed in its interpretation across models (and one naturally observes that doing so presupposes, to some extent, our grasp of ϵ), and why it is so natural to think, when comparing the expressive power of second-order logic with that of first-order logic, one need concern oneself only with the second-order quantifier.

There is an observation to make about the second-order quantifier, however, that has also been overlooked in these discussions. This is that the second-order quantifiers under the standard interpretation not only have to range over *all* the subsets (relations, etc.) relativized to the domain, but that they must also range *only* over such objects. Their ability to do this is helped by two peculiar facts about second-order logic and its model theory: (1) they quantify into predicate places, and (2) their range is only *implicitly given* by the model theory – that is, what they range over appears nowhere in the domain. But

---

24 This is the position sketched in Shapiro 1985, 1990, although he is willing in the latter paper to regard the debate between the first-order theorist and the second-order theorist as a (skeptical) standoff. Hellman (1989, p. 69) writes, 'The slogan would be, " 'All' means *All*." ' On the other hand, Putnam (1977, p. 23) complains that ". . . the 'intended' interpretation of the second-order formalism is not fixed by the use of the formalism (the formalism itself admits so-called 'Henkin models,' i.e., models in which the second-order variables fail to range over the *full* power set of the universe of individuals), and it becomes necessary to attribute to the mind special powers of 'grasping second-order notions'."

25 Subject to the qualifications mentioned five paragraphs back.

again, when we translate the language into truncated first-order logic, what is obvious is that the range of the first-order quantifiers, which mimic the second-order quantifiers under the standard interpretation, must be fixed by predicates that hold over *part* of the domain. Somehow, we have been given the capacity to focus on part of what there is to the exclusion of other things there are (relative to the model, of course).

The fact is, second-order logic with standard semantics is *treacherous:* Its sirenlike notation can lull philosophers into an inadequate appreciation of how it gains its expressive powers. For if philosophers are not willing to accept truncated first-order logic as a solution to the problem of referential access (and what I have cited of Shapiro certainly suggests he shouldn't be so willing), then the equivalence given earlier shows that they shouldn't accept second-order logic with standard semantics, or more powerful alternatives as solutions, either.

The conclusion to draw, from the considerations given in this section and in Section 2, is that ontologically conservative strategies will not help us to solve the puzzle of referential access. In particular, we should not look to higher logics for a solution, nor, as I have said, should we simply tuck the capacity to make such mathematical distinctions into the mathematical employment of natural language – not without an explanation of what it is about natural language that gives us such powers. We must look elsewhere for a solution. We must, somehow, undercut the assumptions behind the puzzle of referential access. And so I turn in the next several sections to motivating such a move.

§4   Intuitions about Reference and Axiom Systems

Clues about the semantic mechanisms presupposed in a practice are often revealed by what sorts of referential mishaps are and are not possible. Recall, as an example, Kripke's (1980) Gödel–Schmidt case. That significant misinformation about Gödel is compatible with successful reference to Gödel via the proper name 'Gödel' was used by Kripke to show that (certain versions of) the description theory of proper names could not be a correct account of how the term 'Gödel' refers.

I think this sort of method can also be applied with philosophical profit to mathematical terms. My main purpose in doing so is to shift the focus of concern away from the puzzle of referential access and more directly on the actual semantic intuitions users of mathematical terms have. Ultimately, such a refocusing will point to a method of undercutting the presuppositions of the puzzle of referential access.

But the intuitions are of interest in their own right – they are data – and any philosopher of mathematics is required, I think, to explain their source somehow.

I should add that although some of the intuitions will be familiar from the literature, most of them will not. A systematic attempt just to describe such intuitions has never been made, in part because the puzzle of referential access (already discussed) and Benacerraf's puzzle (discussed in Section 8) have been so salient to philosophers.

Let's start by repeating the already familiar observation that it is not uncommon to think that the references of mathematical terms are fixed by axiom systems, where the primitive symbols of the language are given their meaning implicitly. In Peano arithmetic, for example, the references of the numerals are fixed by their definitions in terms of the constant symbol '0' and the successor function symbol 's'; the references of these primitive terms are fixed in turn by the axioms of Peano arithmetic. Indeed, as we've seen, one strong motive for choosing second-order Peano arithmetic over its first-order rival is its reference-fixing powers: It is categorical.

Two caveats: First of all, this picture requires a background of set theory to draw the models from, for it is in that context that categorical systems *rule out* alternatives that are not isomorphic to the intended model. Second, given such a background, the categoricity of an axiom system must be sharply distinguished from its completeness, as Corcoran (1980) points out. What is at issue (referentially) with mathematical theories studied against a set-theoretic background is categoricity, not completeness.

However, categoricity becomes irrelevant when we do not have set theory (formal or otherwise) to rely on. For example, our recognition of alternative set theories, at least when interpreted at face value (and only when so interpreted are they *alternatives*), must be compared without such a set-theoretic background. In such a case, we can distinguish set theories, perhaps as frameworks for mathematical practice generally, only via the *different* axioms that hold of them.[26]

26 The contemporary notion of categoricity turns on the fact that the units of significance (sentences) are not wedded to any particular model, but may be reinterpreted at will (subject to certain restraints, of course). The class of models they are interpreted in are drawn from a set-theoretical background, as just mentioned. Corcoran (1981) points out that early postulate theorists such as Dedekind, Peano, Hilbert, Huntington, and Veblen were not working in any formal set-theoretic framework. Corcoran, therefore, defines a second notion of categoricity that would have been available to these theorists. For this purpose he calls a "system" a "complex of objects, functions, relations, etc. and *not* . . . a complex of propositions, symbols, etc." (1981, p. 113). Regard a mathematical proposition about a given system as a Russellian singular proposition containing the elements of the system it is about. Then such a proposition is *not* open to reinterpretation in the style of contemporary model theory. However, one can consider *translations* of these proposi-

With these caveats in place, I can now make my claim: Mathematical practice, and the semantic intuitions accompanying the use of mathematical language, pose grave difficulties for the suggestion that axiom systems fix the reference of mathematical terms. Here are the considerations.

We believe that mathematicians living before the rise of the study of formal languages and formal systems, and in particular, before the invention of set theory, were studying mathematical objects, and in many cases, the same mathematical objects we currently study. Consider, for example, the following quotation regarding the ancient Egyptian method of calculating 12 × 12:

Four times 12 and eight times 12 are added to produce 12 times 12. The numbers which are to be added, are indicated by an inclined line on the right . . . [.] The result 144 is accompanied by the hieroglyph dmd, which represents a scroll with a seal. (van der Waerden 1963, p. 18)

Here it is assumed, it seems clear, that certain Egyptian symbols are taken to refer to certain numbers of ours: 12 and 144 are singled out in particular. It also seems clear that the author not only understands certain Egyptian symbols to stand for certain of our numbers, but that he also thinks Egyptians referred to certain contemporary operations that we refer to by means of the terms 'addition', 'multiplication', and 'division'. Here is another quote that clinches the latter suggestion:

Division is considered by the Egyptians as a kind of multiplication, but formulated inversely. "Multiply 80 (or literally: add beginning with 80) until you get 1120", says No. 69 in the Rhind papyrus, and the solution looks exactly like a multiplication. (van der Waerden 1963, p. 19)

Here he clearly intends to translate some phrase or other in the ancient Egyptian as 'multiply' (or, literally, 'add beginning with').

This identification of particular mathematical objects (and operations defined on those objects) studied or used by mathematically primitive practitioners with contemporary mathematical objects (and operations) is routine among historians of mathematics, and quota-

---

tions from the context of one system S' to another system S', where such translations are gotten by replacing the constituents from S in the proposition with appropriate constituents from S' (I forgo giving precise definitions).

Corcoran then goes on to define this second notion of categoricity in these terms: Let P be a set of propositions about a system S. Let P' be any arbitrary translation of P gotten by substituting the elements of some system S' for S. Then P is categorical if, whenever P and P' are both sets of true propositions, an isomorphism exists between S and S'. Although this second notion does not rely on an explicitly formal set-theoretic background, it does rely on an implicit set-theoretic background (witness the assumption that one can define mappings from one system to another). Therefore I doubt its relevance either to mathematical contexts much before Dedekind or to a context where one is, say, comparing alternative set theories. Of course this is not to imply that Corcoran suggests otherwise.

tions of this sort may be found in just about any discussion one looks at.[27] Here is another. In the following, Smith wrote of the number names of the Andamans, a "tribe of Oceanic negritos." He said:

> Their number names are limited to one and two, but they are able to reach ten by this process: the nose is tapped with the finger tips of either hand, beginning with one of the little fingers, the person saying "one" (*úbatúl*), "two" (*íkpór*), and then repeating with each successive tap the word "*anká*," which means "and this." When the second hand is finished, the two hands are brought together to signify 5 + 5, and the word "all" (*ardùru*) is spoken. (1958, p. 7)

It seems clear that Smith thought that it was the number 5 + 5 (or 10) that was referred to (signified) by the Andamans in this round-about way. It is equally clear, I think, that neither the Andamans nor the Egyptians (nor, for that matter, the more sophisticated Babylonians) had set theory to serve as a background for a categorical set of axioms. Worse, they didn't even have anything much resembling axioms at all, but at best a few rules of thumb, most likely empirically arrived at.[28] In particular, the Andamans (at this stage anyway) didn't seem to have anything resembling an unrestricted successor function.

In saying this, I don't mean merely to point to the fact that the Andamans didn't have a way of generating an open-ended set of (oral) *numerals*. Rather, it seems unlikely that the Andamans realized that numbers go on and on. Nevertheless, there seems to be no doubt (in Smith's mind) that the Andamans did refer successfully to certain small numbers.

Similarly, it is suggested by van der Waerden in chapter 2 of his 1963 book that the Egyptians studied the circle just as the Babylonians did, and that they had an approximation of π that was actually better than that available to the mathematically more sophisticated Babylonians. In other words, both the Egyptians and the Babylonians are taken to have been studying geometric objects without an axiom system that picked these things out.[29]

27 See, as examples, Grattan-Guinness 1970, Kline 1972, or Crowe 1988. Kitcher 1984 constitutes a remarkable, self-conscious exception. I discuss his case later.

28 There is no reason to think that either the Egyptians or the Babylonians knew the absolutely crucial principle of induction. See Kline 1972 or Van der Waerden 1963. This is no surprise. What is immediately salient about numbers is that one can count with them, add them, and multiply them. Recognition of the principle of induction comes later.

29 Also see Kline 1972. Notice it is not reasonable to suggest that van der Waerden thinks the Egyptians and the Babylonians were studying a class of empirical objects (such as sketches in sand, drawings, or three-dimensional physical objects), despite the fact that their formulas for calculating areas and volumes of certain figures were clearly empirical approximations that were in many cases wrong; for the question of, for example, how good an approximation for π they had makes no sense if π is taken to be such an object.

What *was* available to Egyptians and Babylonians (not to mention the Andamans and the Pitta-Pitta of Queensland) for fixing the references of their mathematical terms? Only the terms themselves and certain practices.[30] The latter were practices both with the terminology (rules of thumb – however meager – about how to compute numbers, some practices manipulating geometric diagrams) and nonmathematical applications of the results of such computations. And *this* is taken as sufficient to fix the references of their terms – and, furthermore, to contemporary mathematical objects *we* study, for example, numbers and circles.

As it turns out, the absence of axiomatic means for fixing the references of mathematical terms is not uncommon even in quite mature mathematics. Until relatively recently (given how long the history of mathematics is), there was no set theory to function as a background for categorical theories. But often (after a while) there was a fairly rich body of mathematical principles known to hold of the objects studied, although rarely enough to distinguish what we take mathematicians in those days to have referred to from the several mathematical alternatives we are currently able to produce. Nevertheless, mathematicians are often taken to have referred to such objects right from the onset of their "discovery."

The history of integration serves as an interesting illustration of this common mathematical ability to refer successfully to a class of objects before the "means" (according to the axiomatic picture) are available to do so. Integration is an operation that was (and still is) steadily applied to a wider and wider collection of functions (and functionlike objects such as functionals), usually before the terminology for such things is rigorously defined (according to informal mathematical standards!).

Consider for example, infinite series (in the hands of Newton, Leibniz, and their successors), Fourier series, and the Dirac delta function notation.[31] Such representations were manipulated analyti-

---

30 I am being vague about what sort of terms are pertinent here because it is debatable which grammatical categories number terms belong in. For example, there is fairly good linguistic evidence that (in English) some of our number terms function like quantifiers. See Benacerraf 1965 and the classic Barwise and Cooper 1981. It is also not unnatural to treat them in other contexts as property terms (predicates) and, of course, as singular terms. Furthermore, it could be that different uses of numerals correspond in natural language to different parts of speech, and perhaps further variation is available in different natural languages. My discussion, so far, is independent of these issues.

31 The sort of "nonrigorous" description of how such a function operates in the context of integration that was available before the introduction of the theory of distributions may be found in Condon and Morse 1929, p. 39. They write: "Now evidently the value of

$$\int f(v)\delta(x-v)dv$$

cally for quite a while without sufficient mathematical understanding of how this was possible, where "without sufficient mathematical understanding" means in part the absence of definitions linking manipulation of the terminology with already existing mathematical notation. Unsurprisingly, not all notation survived the subsequent refinements that eventually led to definitions – a fate suffered, for example, by many of the divergent series utilized by Leibniz and Euler.

But the uses of notation *on the part of pioneer mathematicians* that survive subsequent definitions are taken to refer successfully to the objects we *currently* take them to refer to, even when the definitions reconstrue the objects referred to in a way entirely different from the way pioneer mathematicians saw them. For example, we do not assume that Dirac did *not* successfully refer to the function that carries his name, despite the fact that it has turned out not to be a function defined on the real numbers. To put the matter crudely, we are not attaching his name to a piece of notation that was meaningless when he used it.

Furthermore, in the cases where the new terminology successfully picks out old objects already referred to by preexisting notation (such as infinite sums or definite integrals which pick out familiar real numbers), the object is considered to have been successfully picked out at the time the new terminology is introduced even if the means (the proof) that the new representation does the job is subsequently rejected, and even if the terminology isn't given a "rigorous" foundation (according to the requirements of informal mathematics) until long after.[32]

Complex numbers offer a similar picture.[33] The complex numbers apparently picked out by Renaissance notation ('$\sqrt{-1}$', for example)

will be simply $f(x)$ if $f(x)$ is continuous, for the integrand will vanish everywhere except at $v = x$ and here it will suffice to replace $f(v)$ by its value at $v = x$ and take $f(x)$ out from under the integral sign." Eventually, S. L. Sobolev and then L. Schwartz supplied the necessary rigor. See Schwartz 1950. To see distributions in their natural (textbook) mathematical context, see Yosida 1978, Reed and Simon 1972, or Rudin 1973 (the latter has a nice bibliography in its appendix B).

On Euler's infinite series, Fourier series, and on the early development of the calculus generally, see Grattan-Guinness 1970, Kline 1972, and Kitcher 1984. I should add that historical work, as is no surprise, is often controversial, and historians sometimes make mistakes (e.g., see Grattan-Guinness's 1973 review of Kline 1972). But the historical points *I* make in this book about mathematics are elementary enough to be ones that most historians agree on.

32 Philosophers may think that because the rigor afforded by set theory and formal logic is now in place, the sort of phenomenon I am describing is a thing of the past. On the contrary: Consider the current mathematical status of the Feynman integral. A nice description can be found in Atiyah 1990, pp. 70–2.

33 See Nagel 1935 and Crowe 1988. Also see Kitcher 1984, especially pp. 170–7, where a view about what fixes the reference of mathematical terms is given that is in opposition to many of the intuitions I have been describing here and will be describing in the next two sections. (See the end of this section for further discussion of this.)

are taken to have been successfully referred to by mathematicians as soon as they started using them, even though (correct) axioms (and definitions) governing these things were late in coming. Indeed, set theory itself presents the same picture, certainly as that subject was practiced by Cantor.[34]

There is also anecdotal evidence available that seems to illustrate the mathematician's capacity to refer to a type of mathematical object independently of his or her capacity to stipulate categorical axioms, or indeed, any axioms at all. Consider Ramanujan. It is well known in the mathematical folklore that this brilliant number theorist seemed to have only a "shadowy" notion of proof. But for similar reasons he had to lack a grasp of axiom systems and consequently a grasp of categorical axiom systems. He is taken to have successfully referred to numbers anyway. The same may be said of many child prodigies in mathematics, not to mention idiots savants.[35]

Successful reference in this sort of context also occurs among contemporary set theorists and logicians who, perhaps because of the historical proximity of paradox, are quite self-conscious in their mathematical practice; and here the case should sound quite familiar. Consider the axioms of set theory. It is well known that the Continuum Hypothesis is independent of our current axioms. And set theorists are aware that some mathematician may discover new axioms that set theorists will find appealing and that will decide outstanding problems such as this hypothesis.

What is striking is that if this happens, it will *not* be assumed that set theorists today were not referring to sets (i.e., the objects that we will call sets in the future: the ones obeying the Continuum Hypothesis, among other things, if that is how the new axioms go), nor will it be assumed that they were referring ambiguously to sets that obey the Continuum Hypothesis and sets* that do not. Indeed, if this were to be our future view, it would now be our *current* view, and it's not. Rather, the way matters are usually described is to treat the issue of whether the Continuum Hypothesis is true or not as still open.[36]

---

34 Maddy (1990, chapter 2) makes this point about set theory.

35 One might hope to hold on to the "description theory" for numbers by suggesting that individuals who cannot produce the desired axiom system defer their reference to those who can. The considerations already raised suggest this move is hopeless, but in any case surely this was false of Ramanujan, whose grasp of numbers in some ways surpassed that of anyone around him to whom he could have deferred referentially. On Ramanujan, see Snow's introduction in Hardy 1967; see also Newman 1956.

36 I am assuming for the sake of this example that advances in set theory would be supplementary. They might not be and then the case changes. Because set theory is so explicitly formalized, any nonsupplementary change in set theory would have to be regarded as a discarding of current set theory. Some philosophers draw the

So far we have the following. Mathematical terminology intuitively succeeds in referring even if the (informal) definitions and axioms the terms are apparently operating with respect to cannot underwrite their references, or even if axioms or definitions are unavailable.[37]

But an interesting observation must be made about the case where an axiom system or a fairly rich informal context of "truths" *is* available for certain mathematical terms. This is that even in such a case the group of "truths" and definitions cannot be taken to fix the references of such terms. This is because such "truths" and definitions can fail to be *right,* and yet we often still take mathematicians using these terms with (partially incorrect) definitions and truths to refer successfully to what we currently take those terms to refer to. The most dramatic case I can think of is Frege's. Notoriously, the set-theoretic system he was working in is inconsistent. Nevertheless, we take him to have succeeded in referring to the numbers, functions, and (some) sets referred to in his system.

Let me elaborate this example. Frege hoped to define the numbers in terms of a certain set-theoretic construction. His particular construction is not available in certain paradox-free set theories, notably ZFC. Nevertheless it is not assumed that Frege was not referring to numbers (but, rather, to nothing at all) when he talked about them. Similarly, he is taken to have successfully referred to the null set, the power-set operation, and many other sets, and operations on them, despite the inconsistency of his own set-theoretic system.

More benign examples can be found in the history of mathematics, a subject littered with false definitions and false "truths." Consider differentiation. The calculus created puzzlement right from its inception, for definitions of the differential of a function seemed to require infinitesimals, and no one seemed able to supply a consistent set of rules governing these things. Nevertheless, it is not assumed that this undercut the capacity of mathematicians to refer to the functions picked out by the differential notation (e.g., via differential equa-

---

conclusion that in such a case we would recognize that set-theoretic terms as we currently use them don't refer to anything. Quine would take this view, and so would Kitcher (I think). I disagree. See Part II.

37 There are differences I am papering over (for the moment) between intuitions about singular terms in mathematics and intuitions about kind terms in mathematics, which turn on the fact that kind terms refer to collections. Thus, although the remarks I have made about successful reference being possible even in the absence of (correct) definitions or axioms intuitively hold of both kind terms and singular terms, still, it is not uncommon to observe that a kind term (such as 'function' or 'real number') does not pick out (nowadays) quite the same things as it picked out in the past; see, for example, Grattan-Guinness 1970, Kline 1972, Lakatos 1976a. Notice that such remarks about changes in kind terms, although couched in intensional talk about concepts, clearly are meant to imply changes in the extensions of such terms. I discuss this more deeply in Section 6.

tions). Similarly, consider the inconsistent rules applied to recognize the summations of infinite sums. As pointed out already, it is not assumed that particular cases of Euler's infinite series, even when convergent in the contemporary sense, failed to pick anything out.[38]

Consider the following. Mathematical papers often have mistakes in them. Proofs and definitions are often fallacious. And by this I mean that the result is correct – that is, the statement shown is (more or less) true; the mathematical object defined to exist does exist – but the proof or definition given doesn't do the job, and sometimes the error is substantial enough that some other mathematician gets significant credit for the repair. Nevertheless it is not assumed that the first mathematician *failed to refer* to the objects whose existence he tried to show or define.[39]

Let me sum up the points I am pressing here. The problem with the suggestion that the references of mathematical terms are fixed by definitions in axiom systems, or implicitly as primitives in such systems, is that it doesn't allow for either substantial errors or more rudimentary mathematical practices where such axiom systems are missing. Even contemporary mathematicians often make mistakes and labor under such mistakes for quite some time. And yet we do not take their terms to fail to refer to whatever it is they are studying.

Formal systems and the definitions needed to define mathematical objects in terms of such formal systems emerge at the end of a practice and codify that practice[40] – but, as we've seen, the terms involved are taken to refer long before that event. Indeed, the process seems to be like this: One can introduce new objects by extending some mathematical operation in some way (e.g., the square root operation to negative numbers) or by introducing new operations (e.g., differentiation). Then one labors to derive results extending other mathematical operations to such objects or to derive results about how the new operations interact with the old ones (e.g., the multiplication rule for differentiation). Finally, *if it is possible,* one uses the body of results thus laboriously gained to define the extension in terms of already existing mathematics.

Notice that the last move is *not* always possible, and its failure need not imply any difficulties for the new mathematics.[41]

---

38 For elaboration of these examples, see Grattan-Guinness 1970, Kitcher 1984, and especially Kline 1972, chapters 17, 19, and 20.

39 For examples of such errors, see Grattan-Guinness 1970, Kitcher 1984, or Lakatos 1976a. Also see the fourth misconception about the history of mathematics in Crowe 1988.

40 See the fifth misconception about the history of mathematics in Crowe 1988.

41 For example, analysis cannot be founded on number theory as Kronecker hoped, but only on number theory plus set theory.

Three caveats: First, one might try to avoid the conclusion that axiom systems do not fix the references of mathematical terms by arguing that the references of such terms are implicitly fixed by the *subsequent* mathematical definitions. I am not entirely unsympathetic with this move, but it simply will not do. The problem is that it seems clear that pioneer mathematical practice does not operate with such definitions implicitly in mind.

Compare the grammatical rules of English (whatever *they* are) and the convergence tests for recognizing well-defined infinite series. Native use of English can be regarded as implicitly governed by English grammar because one can (in theory, anyway) explain violations of that grammar as a *function of* performance factors, for example, laziness, memory failures, and the like, *and* the grammar itself. English grammar thus plays a role in explaining what native speakers are doing even when they are being ungrammatical. It is this fact that enables us to describe English grammar as implicitly at work in native speech, even if the speakers *don't know* that grammar (i.e., can't cite the rules).

But there seems nothing analogous to explain the formal use of divergent series by eighteenth-century mathematicians. The combination of formal manipulation plus the pragmatic avoidance of problematical cases suffices to explain their practice. There is no need to presume, and no way to justify, a logical deep structure in their mathematical practices that presupposes, say, the advances made by Borel.

It may seem that historians of mathematics will disagree. For example, Kline writes:

The theory of asymptotic series, whether used for the evaluation of integrals or the approximate solution of differential equations, has been extended vastly in recent years. What is especially worth noting is that the mathematical development shows that the eighteenth- and nineteenth-century men, notably Euler, who perceived the great utility of divergent series and maintained that these series could be used as analytical equivalents of the functions they represented, that is, that operations on the series corresponded to operations on the functions, were on the right track. Even though these men failed to isolate the essential rigorous notion, they saw intuitively and on the basis of results that divergent series were intimately related to the functions they represented.[42] (1972, p. 1109)

But we should distinguish, I think, between glimpses that a theory is available for rigorizing (or explaining) one's practice and actually

---

42 On the other hand, fairness to me requires that I point out that a close reading of other parts of Kline's book shows that he is quite aware that one cannot explain eighteenth-century practice in terms of an intuitive grasp of the "correct" principles governing infinite series. See his chapter 20, in particular, for details describing the loose practices and confusions such mathematicians had regarding infinite series.

implicitly using the theory in one's practice. Arguably, only in the second case can the theory be taken as helping to fix the references of the terms used in practice.

A similar point dashes the hope that one might explain Greek and Babylonian usage in terms of an implicit formal system embodied in *their* natural language. Cardinal quantifiers aside (if there are any operating in natural languages), mathematical practice rapidly outstripped its natural-language context; it is hard to believe that a formal system (or set of systems) in natural language could have been in place that referentially fixed these advances. For the same reason, it is hard to believe that all our current practices with numbers are fixed by implicit terminology already available in contemporary natural languages.

It may seem that I have required too much here. For consider: I have suggested that if one says that what fixes terminology at a time t is a certain axiomatization (which only emerges at some future time $t + n$), then a requirement for this claim is that the axiomatization in question must play an explanatory role in the terminological practices at t.

But why is such an explanatory role required? Why isn't it simply that scientific practice (mathematical practice included) presupposes that the references of the terms used in a science are fixed by whatever fixes the terminology in the mature version of the science – in this case, by a forthcoming axiomatization?[43]

There are two problems with this move. First, mathematical practice went on successfully for thousands of years without any worries about what fixed mathematical terminology. Mathematicians certainly expressed worries about what was true or false of what these terms applied to, but that is a different matter. This suggests that, intuitively, there is nothing free-floating about mathematical reference.

Second, one can certainly imagine that mathematical development might have been forever arrested at a certain primitive point. (Our mathematics might simply have not advanced past what was available to the ancient Egyptians; or we could have become extinct.) Certainly, it doesn't seem as if an alien from outer space watching our lack of progress would have to conclude that our mathematical terms in such a case didn't refer to anything. The only way around this latter objection that I can see is to suggest that mathematical development always tends toward a Peircean ideal (a certain perfect axiomatization) that can then be taken by definition to fix the references of mathematical terms. But Gödel's theorem makes this move extremely implausible.

---

43 Mark Richard raised a version of this objection to me in conversation.

This is why I considered only the stronger version of this approach, which requires that if an axiomatization is taken to fix the reference of terms, then it must be operating implicitly in the practice *at the time the references of the terms are supposedly fixed by this axiomatization.*

The second caveat: Nothing I have argued for so far denies the role of axioms, or mathematical truths, in *distinguishing* our reference to one sort of mathematical object from our reference to another sort. Consider the alternative set theories currently available. There is Quine's *ML*. There is ZFC. There are the various type theories. There are the various species of nonfoundational set theories. And of course, there are intensional set theories, multivalued set theories, fuzzy set theories, and so on. That the terms in such alternative theories refer to different objects is recognized by the different axioms governing them. This can be conceded while still denying that such axioms fix the references of such terms. For pointing out a truth that holds of As but not of Bs will suffice to distinguish As from Bs without sufficing to fix the reference of either 'A' or 'B'.

The final caveat: The considerations raised here offer a prima facie objection to approaches that attempt to replace the reference-fixing capacity of axiom systems with something broader, such as "practices." For example, Kitcher (1984, pp. 163–4) takes a practice to be a quintuple <L, M, Q, R, S>, where L is the language of the practice, M is the set of metamathematical views, Q is the set of accepted questions, R is the set of accepted reasonings, and S is the set of accepted statements.

Now Kitcher is not entirely clear about what individuates such practices, that is, exactly how much in a practice can change before we take the terms in a mathematical language to have shifted in reference. But the examples he gives show that he is much less willing to allow sameness of reference in cases where a practice has shifted than most historians of mathematics.

As an example, consider his position on complex numbers; see Kitcher 1984, pp. 175–6. In the sixteenth century, as mentioned, expressions like '$\sqrt{-1}$' became fairly common. As this was a new use of old language, the question arises, to what did '$\sqrt{-1}$' refer? Kitcher suggests that there were two available ways in medieval and Renaissance practice for mathematicians to fix the reference of the term 'number'. The first was to "use the available paradigms—3, 1, −1, $\sqrt{2}$, π, and so forth—to restrict the referent to the reals" (p. 175). The second was to take numbers as any entities on which arithmetic operations can be performed. The latter paradigm *does* admit complex numbers as numbers as soon as one has notation for referring to them, and it is clear that the new entities will submit to arithmetic operations. Consequently, Kitcher concludes, Renaissance mathematical practice was not definite on the issue of complex numbers and did not become so until "[g]radual recognition of the parallels between complex arithmetic and real arithmetic led to repudiation of the more restrictive mode of reference fixing, so that the reference potential of '$\sqrt{-1}$' came to include only events in which i was identified as the referent" (p. 176).[44]

Intuitively, this view has some of the same drawbacks as its predecessor already discussed: The temptation is almost overwhelming to describe the Renaissance introduction of complex numbers as the introduction of *complex numbers*, regardless of what Renaissance mathematicians thought they were doing (as evinced by their use of such terms as "sophistic," "nonsense," "imaginary," etc.), and this, in fact, is how historians discussing the period actually write about it.[45]

As we've seen, Kitcher includes as part of the practice the metamathematical beliefs that, among other things, mark out which terms are taken to refer and which treated as instrumental heuristics. *But* a mere shift in such attitudes does not cause a shift in which terms refer and which don't. It is hard to see what more is needed of '$\sqrt{-1}$' to refer to i other than the fact that it operates (mathematically) as the name of $\sqrt{-1}$![46]

44 Let me complain a little about Kitcher's description of what happened here. First off, it was not merely complex numbers that were distrusted: Negative numbers were distrusted for nearly as long; indeed, mathematicians would avoid getting them as roots for equations if they could do so by modifying the equation. Second (or so it seems to me), it wasn't so much that mathematicians had to recognize that extending paradigmatic arithmetic practices to such numbers would be successful; rather, it was certain specific confusions or problems that were slow in being eliminated that caused the distrust. In particular, there were problems about how to understand ratios of negative and positive numbers, and there were problems with the ordering of such numbers. Euler, for example, believed that negative numbers were greater than ∞. Indeed, we find that negative and complex numbers were still distrusted by De Morgan; see Kline 1972, p. 593.

45 Kitcher (1984, p. 156) does claim that traditional history of mathematics is tainted by a priorist methodological prejudices, and perhaps it could be argued that these observations about the transpractice reference of (certain) mathematical terms are examples of that. But I don't think so, as such intuitions are compatible with mathematicians getting definitions and theorems *wrong*. What these observations do point to is something of a semantic analogy between mathematical singular terms and kind terms, on the one hand, and empirical singular terms (e.g., proper names) and natural kind terms on the other, something I suspect would not be appealing to an a priorist. I hasten to add that the a priorist should not be upset yet, as we shall soon see that the analogy is strictly limited in its force.

46 Strikingly, Crowe (1988), who is otherwise quite sympathetic to Kitcher's views, cannot help describing complex numbers in ways which presuppose successful reference to them right from the time of Cardan.

   I should add that I am temporarily burying a complication. This is that when one takes complex numbers to be ordered pairs of reals, there are *two* possibilities for the referent of 'i'; that is, there is a residual problem remaining about *which* number (i.e., which pair of reals) the Renaissance mathematicians actually had in mind. I treat the issue this raises later (see Section 6).

Three last observations to wrap up my discussion of Kitcher's views: First, these views about the reference-fixing powers of practices seem to work somewhat better with kind terms than with constants, as we'll see. Second, although Kitcher doesn't worry about our current practices in attributing reference to earlier mathematical terms, he does worry that the suggestion that '$\sqrt{-1}$' refers to i right from the application of the square root notation to '-1' makes the prolonged debate over complex numbers mysterious. But this is false. The prolonged debate can be explained, precisely as Kitcher does, in terms of differing paradigms. One can accept such an explanation while rejecting the claim that such paradigms *fix* the references of mathematical terms.

Third, I don't want to diminish the importance of the worry helping to motivate Kitcher's approach: that *something* must be fixing the references of our mathematical terms. If it isn't definitions and axioms, it isn't practices, and it isn't intellectual intuition, then *what* could be doing it?

But one cannot help adding the following. Concern with reference fixing does not seem to be a *mathematical* worry. When one looks over the history of mathematics and examines the motivations for deductive rigor and definitions, what is obvious is that such motives were (and are) *epistemic:* The mathematicians were (and are) concerned with getting mathematical results and securing their validity. Categoricity arises gradually and late as a topic of study. The current philosophical concern with how mathematical terms pick out what they refer to is an oddity from the point of view of mathematical practice, which, in broad respects, is simply not concerned with reference. Any view that fails to explain why this is the case has not explained something crucial about mathematics.

§5   Comparing Mathematical Terms and Empirical Terms I

The considerations lately raised give the impression that mathematical terms act like empirical natural-kind terms and proper names rather than like terms whose references are fixed by description – at least insofar as substantial error and ignorance about objects and kinds are compatible with successful reference to those objects and kinds.[47] Certain philosophers prone to assimilate mathematics to em-

---

47 As I've already mentioned, the classic discussion of proper names may be found in Kripke 1980. Similar observations about natural-kind terms may be found in Putnam 1975c. I should add, although I will not dwell on it in this book, that natural-language quantifiers seem to come equipped with context-fixed extensions – so they, too, do not seem to be terms that are fixed in their operation by "definitions."

pirical science no doubt will be heartened by such resemblances, and may take these results to license the suggestion that mathematical terms pick out objects to which a causal-style theory of reference can apply. Maddy (1990), for example, claims that sets of empirical objects are perceived in much the way that the empirical objects themselves are perceived, and indeed, are located in much the same place that the empirical objects that are their members are located.

Unfortunately, there are further intuitions, about the contrast between referential errors possible with mathematical terms and referential errors possible with empirical terms, that suggest serious *disanalogies* between these sorts of terms: Any assimilation of the mathematical term to the empirical term seems to face problems almost as difficult as the problems faced by the doctrine that the references of mathematical terms are fixed by axiom systems. Let's start the presentation of this contrast with an attempt to classify the sorts of referential errors that one can make with empirical terms.

Whatever the facts are about how a singular term gets its reference fixed – whether, that is, by a description or by, broadly speaking, causal means – speakers in practice can use either a description or ostension in their attempts to determine what their terms refer to. Consequently, regardless of the semantic status of the singular term being used, there are several ways a speaker can get the reference for that term wrong. I should add that the sorts of mistakes I am going to discuss now are ones that speakers can make without our drawing the conclusion that they are incompetent in the uses of the terms they've made mistakes with. Other mistakes do apparently betray incompetence. We'll discuss those later.

Let us first take up ostension. Ostension admits of two sorts of mishaps, which we'll call primary A- and 'A'-mishaps.[48] Primary A-mishaps are cases where we confuse the referent of a term with another object. Primary 'A'-mishaps are cases where we are not confused about the referent but, rather, about the *term used* and what it refers to. The first sort of error has its source in a confusion that is not essentially connected to the use of language (we can understand the nature of the confusion without bringing in the term at all); the second confusion, however, requires a description of how the misused

It is an interesting question (which I also do not address in this book) whether an intuitively appealing "causal story" of the sort Kripke, Putnam, and others have designed for predicate terms and constants applies to quantifiers.

By way of anticipation of what I will say shortly, one finds Kripke (1980, p. 60) toying with the idea that π functions like a proper name rather than like a defined term. But he does not take the matter very far, and so one doesn't know how serious he is. The suggestion that mathematical-kind terms function like natural-kind terms may be found in Maddy 1990 chapter 2, as already mentioned.

48 Read " 'A'-mishap" as "quote A mishap."

term came to be misused. I go on to give examples of primary A- and primary 'A'-mishaps in the case of proper names.

Suppose John and James are twins. I may see John and think he is James because, quite simply, I am bad at distinguishing twins. This is, intuitively, not a matter of using names in the wrong way. I use 'John' to refer to John, and 'James' to refer to James, and so, when I call John 'James' I am not mistaken about what 'James' refers to. Rather, I am mistaken in that I am attempting to apply the name 'James' to John. This is a *primary A-mishap*.

By contrast, suppose I am still laboring under the misapprehension about the twins, and Sarah asks me who the redhead is. When I tell her he is James, Sarah (who is quite good at distinguishing twins and knows which one she likes) thinks 'James' applies to John (and not James); because of my error, she is mistaken about the reference of the term 'James'. This is a *primary 'A'-mishap* (Sarah is confused about the term 'James', not the object James).

Here is another example of a primary 'A'-mishap. Suppose Sarah's hearing is impaired in such a way that she cannot easily distinguish the sound of 'John' from that of 'James'. Then our second example could have gone this way: I tell her that *his* name is 'John', but alas, she hears 'James', and applies the name 'James' to John as a result.

Let us now illustrate primary A- and 'A'-mishaps in the case of singular terms that are not proper names.

Imagine that John's father has (roughly) the same build as the local blacksmith, and I see the local blacksmith from behind with his arm around a woman who is not John's mother. In this case, if I draw the conclusion that John's father is having an extramarital affair, I have confused John's father with the local blacksmith, a primary A-mishap. Suppose I am walking with Sarah when this happens, and suppose she doesn't know either John's father or the local blacksmith. If I say, "Look, that's John's father with his arm around another woman," Sarah will mistakenly think that the term 'John's father' applies to the local blacksmith. This is an 'A'-mishap. We can also rig the example so that Sarah mistakenly thinks that the term 'the local blacksmith' applies to John's father, for consider this case: We are walking behind John's parents; I think, mistakenly, that John's father is the local blacksmith, and I complain to Sarah, gesturing toward the woman in front of us, that John's mother is having an affair with the local blacksmith.

I mentioned earlier that speakers can attempt to fix the reference of a singular term by means of a description; when this happens, the mishaps possible are more complicated. In order to classify these, it is easier if we first consider the generalization of primary 'A'- and A-mishaps to empirical-kind terms. This generalization, by the way, will prove to be important independently of the issue of what sorts of mishaps are possible with singular terms.

First off, three sorts of primary A-mishaps commonly arise with kind terms. We can mistakenly think that an object b is an A, we can mistakenly think that As are Bs (the two groups are identical), or we can mistakenly think that all As are Bs (any A has the property B).

When I apply the term 'chair' to something, I may get it wrong, for the object may not be a chair at all: It may be a carefully crafted illusion. When it comes to terms that apply to ordinary objects like chairs, this sort of error is always possible.

However, the same is true of theoretical objects, *provided that* the theory of such objects, and the context they operate in, are fairly settled.[49] For example, the theory of electrons seems fairly well established (at least in regard to certain branches of physics), and consequently it is possible for me to misconstrue some phenomenon as involving electrons (perhaps because I think electricity plays a role there) and get it wrong because electrons are not involved in that way at all. If something plays the kind of causal role in the phenomenon that I was attributing to electrons, I commit a mishap of exactly the same sort as in the chair case. Notice that the intuitions allowing this sort of mishap apply both in cases where we take the pertinent term to be a natural-kind term and where we don't. I may think that such and such stuff is gold, although it isn't; but I may also think *that* is a chair, although it isn't, as we've seen.

Illustrations of the other sorts of primary A-mishaps are also easily forthcoming. While studying a group of squirrels, I may come to think that two distinct species of squirrel are identical because they inhabit the same area and look quite alike; this is the second sort of primary A-mishap. And as an example of our last kind of mishap, here's something somewhat fanciful: I may think that a certain sort of ant is carnivorous because it hunts down small rodents in packs, and the pack drags off whatever it kills to its lair. But, in fact, these ants may be using the rodents for some other purpose.

What about primary 'A'-mishaps with kind terms? We can easily manufacture examples along the lines I used to manufacture such mishaps for singular terms. For example, suppose I visit my sister at her lab, see an unusual-looking hamster in a cage, and ask her what it is. Suppose she is not really paying attention to me and she, offhandedly thinking I am pointing to another cage, says it is a squirrel. I will leave with the impression that there is a kind of squirrel that looks very much like a hamster. This is a primary 'A'-mishap. I will continue to pick out successfully that sort of hamster, but I will mistakenly ap-

---

49 See Section 6 for a discussion of what happens when the background theory and context go into an upheaval.

ply the term 'squirrel' to it. Similarly, I may play a cruel joke on someone who is learning English – I may tell him that 'carnivore' applies
not only to animals that eat meat but also to animals that we eat. He
will go on (for a while at least) applying 'carnivore' to sheep and
shellfish.[50]

We return again to singular terms. As I mentioned earlier, sometimes speakers attempt to fix the references of their terms by means
of descriptions they associate with the terms. In these sorts of cases,
we have examples of primary A- and 'A'-mishaps that are a little different from the sorts of cases that arise via ostension. Furthermore,
two other broad classes of mishaps, what I call secondary 'A'- and A-
mishaps, are possible. These can arise because of the semantic structure of the description being used by the speaker to fix the reference
of the term at issue; for primary A- and 'A'-mishaps with respect to
the terms occurring in that description can derivatively cause mishaps
with respect to the term the description is associated with. Also, secondary A-mishaps for a term can arise because of primary A-mishaps
with respect to collateral information pertinent to the description being used to fix the reference of that term; secondary 'A'-mishaps for
a term can arise because primary 'A'-mishaps with respect to other
terms can cause mishaps with the term at issue. Illustrations of all of
these sorts of mishaps follow.

Let's start with an example of a primary 'A'-mishap that arises by
means of an associated description. Suppose Samuel overhears a conversation between two officemates where something named Dalila is
described as a "blond bombshell." As it turns out they are discussing
a poodle that a somewhat silly friend of theirs has just dyed; unfortunately there is a woman Samuel thinks this description fits (and
suppose he is right). His subsequent assumption that her name is
Dalila is a primary 'A'-mishap: He is not confusing the woman with a
poodle (or with anything else, for that matter), nor is he mistaken
about any of her properties; he is confused about what the term
'Dalila' applies to.

On the other hand, suppose Dalila is described by the officemates
as having a secret lover. Imagine, further, that Samuel has noticed a
certain woman in the office making phone calls on the sly. However, unbeknown to him, this is because the woman has been using

50 One other sort of primary 'A'-mishap in regard to kind terms is still available in
logical space, although I'm not sure it ever arises: It is one that involves not kinds
but the application of a kind term to an individual. Suppose I tell someone learning
English that the term 'carnivore' applies to animals that eat meat, plants that eat
insects, and my orchid. If his native language is similar to English, the example
won't work, I think, because either the fellow will assume that I mean that the term
applies to *all* orchids, or he will assume that I am pulling his leg. But, perhaps, if he
is sufficiently gullible about foreign languages, he could believe me.

the office phone to set up job interviews. Here Samuel's error is an A-mishap: Samuel has mistaken getting-a-job-interview-on-the-sly-behavior with talking-to-a-secret-lover-on-the-sly-behavior. Notice that Samuel's confusion is not between Dalila and someone else but between one *event* and another, where the events in question are pertinent to the description associated with the name 'Dalila'. The error, therefore, is a *secondary* A-mishap.

Here is another case. Suppose I meet a man called Dirk, and I am quite struck by the peculiar shape of his nose, so struck in fact, that years later, I can only really recall that peculiar nose. I see someone that has the same-shaped nose and draw the conclusion that he is Dirk. Here, it seems to me, I commit a primary A-mishap: I have confused Dirk with someone else on the basis of *properties* that apply to both of them.

Two points: Notice that in both this case and the first Dalila case we considered, the properties described play a role in the mishap (in contrast with the blacksmith cases, where the definite description is being used "referentially" in the sense of Donnellan 1966). But the roles they play are significantly different ones. In the Dalila case, the properties associated with the term 'Dalila' are used to pick something out in order to apply the term 'Dalila' to it. It does not seem reasonable to say that Samuel has confused a woman with a poodle. On the other hand, this is exactly how we do describe the case of Dirk. I have confused two men on the basis of their noses.

It may seem that the John/James case we first considered should be seen as the same sort of case as the Dirk case. In picking out the wrong twin, isn't it the case that I have fastened on some property I take one twin to have and the other not to? Well, this *could be* how the case of the twins works, but in fact, it need not be. One may simply be rather bad at distinguishing twins – two objects may be confused without it being the case that there are *properties* that one had in mind to distinguish the objects with.

Here is another secondary A-mishap. Imagine that I am introduced to John's father at his home and see that he has a number of blacksmithing tools that he is obviously quite familiar with. I may hastily draw the conclusion that John's father *is* the local blacksmith, and be wrong (blacksmithing is just a hobby for John's father; he actually raises sheep for a living). Here, I think John's father has the property of being a local blacksmith and he doesn't.[51]

What about secondary 'A'-mishaps? Here's one: Imagine that Sarah is still laboring under the primary 'A'-mishap that 'James' refers

---

51 Notice that here the A-mishap is not a matter of confusing one object with another but, rather, a matter of thinking an object has properties it doesn't: It is an example of the first sort of A-mishap involving kind terms that we discussed.

to John. Suppose a woman is pointed out to her with the phrase, "This is his fiancée," where in the context it is made clear that 'his' refers to the man she thinks is James. She will subsequently misapply the term 'James' fiancée' to John's fiancée, a secondary 'A'-mishap.

There is a large class of primary and secondary 'A'-mishaps that I have not yet discussed. As it turns out, this class is relatively rare for empirical singular terms, although rather common when it comes to mathematical singular terms. In an obvious sense much of mathematics involves the syntactic manipulation of terms, and errors in reference can arise because of this. Such sorts of errors are uncommon when it comes to empirical terms, but here are some stabs at examples.

Imagine that Edward, a high school student, is copying out passages from an encyclopedia for his homework. In one passage he is copying, a single sentence with the term 'blacksmith' occurs: "Johann Friedrich Gude, a blacksmith originally from Wittenberg, was a popular figure in the early 1600s"; but imagine that he miscopies it as: "Johann Friedrich Gude, a blackfoot originally from Wittenberg, was a popular figure in the early 1600s." When he looks over his paper later, he is struck by the fact that there were Blackfoot Indians in seventeenth-century Wittenberg. Was there a reservation there? he wonders. And he also asks himself: Why did one of them adopt a German name? This is a secondary 'A'-mishap.

Here's another example of a secondary 'A'-mishap. Imagine I am tracing back my family tree. In trying to determine exactly how one of my ancestors, Hinrich Föge, is related to me, I leave out a 'great' and refer to him (wrongly) as my great-great-great-great grandfather. By contrast, imagine this case: Hinrich Föge was ashamed of his son's marriage to a woman from a poor family. When tragedy struck (the young parents were killed in a hideous cart accident), he adopted the grandchild but managed, by means of an elaborate hoax, to fool the community into thinking the child was his son. This myth was passed down through the family so successfully that I am under the impression (wrongly) that Hinrich Föge is my great-great-great-great grandfather. This is a secondary A-mishap.

I mentioned earlier that one can make errors that are prima facie evidence of linguistic incompetence, and I should really give some examples of these too. Nonfanciful cases are hard to come by with proper names, so here's a fanciful one. Suppose we realize that Mike thinks James is called 'John' (as well as 'James') because he thinks that if a male has a proper name beginning with 'J' then he has *every* proper name beginning with 'J'. In such a case we would conclude that Mike doesn't know how to use proper names, he is *linguistically incompetent* (relative to proper names). Such examples of linguistic incompetence are far more common with definite descriptions and other singular terms, because the internal semantic structure present in definite descriptions allows play for such errors. If I am mistaken about what the term 'blacksmith' means, I may think John's father is a blacksmith because he shines shoes. This is linguistic incompetence, at least in regard to the term 'blacksmith'.

Let me stress again that, regardless of the actual semantic status of an empirical singular term – that is, regardless of whether it is a proper name, a definite description, or whatever – we find that use of such a term is susceptible to primary A-mishaps. I especially want to stress this point because the susceptibility to primary A-mishaps on the part of a set of singular terms seems to be a good indicator that our referential access to what the terms refer to is open (in such cases) to a causal story, or at least to a story that does not *require* that our access to what such terms refer to be mediated via a description. Interestingly, this can be the case even when the semantics of a term requires that its referent *is* picked out via a description, as we've seen.

I should say that the classification scheme I've given may seem a little hard to apply in certain cases. I have described the Dirk case as a primary A-mishap, but perhaps one can see it as a secondary A-mishap – because the mishap is due to a property Dirk shares with someone else rather than being due directly to a confusion of Dirk with someone else. One may find the mishap more analogous, that is, to the mishap I commit in the case where I meet John's father at home amid his blacksmithing tools than to the mishap committed in the case of the twins. My own thinking is that in the Dirk case, there has been *no* mishap with properties but only with Dirk, and this is why the mishap in question is not a secondary A-mishap.

There may also seem to be a certain fuzziness with the distinction between primary 'A'-mishaps and secondary 'A'-mishaps in the case of descriptions. For a 'A'-mishap with the term 'blacksmith' in the case of the term 'the local blacksmith' could be seen as secondary or primary, depending on whether mishaps with the proper semantic parts of a term are secondary or primary with respect to that term. Nothing will turn on how we refine the distinction between primary and secondary mishaps in order to legislate on this issue, so I won't bother. I certainly don't want to claim, in general, that I am always going to be able to produce a story that will sort every example one can come up with into one or another category that I have described. All I can hope for is that the distinctions, as they stand, are clear enough to allow us to see in what ways referential errors possible with empirical terms differ from those possible with mathematical terms.

With this in mind, we turn to singular terms in the mathematical case. To start with, it seems that primary A-mishaps are *not* possible

with mathematical singular terms that refer to small whole numbers.[52] Suppose, for example, Jane makes a mistake in integration
and concludes that

$$\int_0^{2\pi} 2\sin\frac{\theta}{2}\, d\theta = 2$$

Intuitively, this is not describable as a case where Jane has confused
the number 4 with the number 2. Well, then, what sort of error has
she committed? My thinking is that it is either a primary or secondary
'A'-mishap. For she thinks the term

$$\int_0^{2\pi} 2\sin\frac{\theta}{2}\, d\theta$$

refers to 2.

Someone might argue that there is a possibility of a secondary A-
mishap here. Why not, that is, say that her error is that she thinks 2
has the property of being the area under the function $f(\theta) = 2\sin\theta/2$,
from 0 to 2π, and she's wrong? The way to dismiss this intuition is to
develop the example far enough so that we see exactly how the error
arose. Generally, how one manipulates a problem of this type is to
generate a series of terms and/or equations, each of which is gotten
from the one before it by a syntactic transformation of some sort. One
kind of error is the simple matter of miscopying. Suppose Jane pro-
ceeded this way: First, she looked up '∫sin $ax\, dx$' in a table of integrals.
Then she substituted 'θ' for 'x', '1/2' for 'a', in '$-1/a \cos ax$', and finally
evaluated the result. Several sorts of errors are possible. She may have
simply failed to include the '2' in front of '2sin $\theta/2$'; she may have re-
membered the '2' but overlooked it when she evaluated the integral;
she may have overlooked the '2' that arises from '$\theta/2$' when integrat-
ing. Or perhaps she made any number of other similar errors of this
simple typographical sort. Notice, though, that we don't want to say
that Jane thinks the function $f(\theta) = 2\sin\theta/2$ is the function $f(\theta) = \sin$
$\theta/2$, for that would be out-and-out incompetence, and her error
doesn't deserve that interpretation. The error is definitely a secondary 'A'-mishap, though which, precisely, turns on how we develop the
example.

Similar remarks apply to Jane's conclusions that $7 \times 8 = 49$, and
$310 \div 5 = 61$. How might these errors arise? Suppose that in the first
case, Jane used the algorithm of adding 7 eight times, and she simply
miscounted how many '7's she used. In this case, we don't want to

52 For the sake of discussion here, I am taking numerals to be singular terms, at least
when one operates with them computationally. For a discussion of intuitions about
kind terms in mathematics, see Section 6.

say that she thought, even momentarily, that 7 follows 5. Similarly, if she simply miscounted the number of times that 5 "goes into" 310, we don't want to say that she, even momentarily, thought that 13 follows 11.[53]

Notice that my initial statement of what I am trying to establish here seems weak: Our examples seem to illustrate that both primary and *secondary* A-mishaps are impossible to make with mathematical terms. I avoid the stronger statement, however, because I think that our intuitions do allow secondary A-mishaps, especially in the case where the error is due to a systematic misunderstanding of a mathematical phenomenon. See Section 6.

In any case, intuitions to the effect that primary A-mishaps are not possible are even more striking when it comes to numbers in canonical notation. Imagine that Eddie refuses to divide by 7, and claims that when 7 is added to any number n, the result is n. I don't think it is possible to conclude that he is referring to 0 but is laboring under the misapprehension that the number 0 *is* 7. Rather, he is confused about the term '7'. We may even be prone to think here not that we are in the presence of an 'A'-mishap but, rather, that Eddie is linguistically incompetent (relative to number terms, anyway).

The same seems to be true when numbers are referred to by non-rigid descriptions.[54] Not long ago, people thought the number of planets was 8. Intuitively, this is not a case where people confused 8 with 9. Rather, it is a case where they mistakenly thought 'the number of planets' refers to 8.[55] As evidence for this intuition, notice that when the error was corrected (which happened when Pluto was discovered) it was *not* described as one in which the confusion of 8 with 9 was finally sorted out. This is a 'A'-mishap.

I am going to call this our capacity for infallible A-attributions (or "A-infallibility," for short) to (small) whole numbers. And, as the integral examples make clear, such a capacity extends to new notation for these numbers.

---

53 How are "miscountings" possible? Well, here's how *I* do it, anyhow. When I am writing a sentence, should I happen to say a word in that sentence aloud, I will skip the word and go on to write the next one instead (the result, of course, is a typo). One can easily imagine someone doing the same thing when counting. Anyway, something analogous is usually involved when syntactic errors of this sort are made.
    One other point: In the example where Jane divided 310 by 5 and got 61, one can see the error happening even if she used short division. She put '6' down, and then (perhaps because of some effect of the numeral '1' occurring in '310') put '1' down instead of '2'.

54 See Kripke 1980, p. 48 for a discussion of rigidity.

55 Why isn't the error a secondary A-mishap in that people thought that 8 had a property it doesn't have, namely, being the number of planets? The problem, I think, is that the error here doesn't seem to be one about numbers and their properties at all.

What about large numbers? A-infallibility *seems* to lapse. Suppose, based on a number figure she read in the newspaper, Marsha thinks that the federal deficit is over 3 trillion dollars (suppose she miscounted the zero numerals). When I first present this example to people, many see it as a case where Marsha *has* confused 3 trillion with 300 billion.

What about other numbers? Well, if someone is given the assignment of reducing the fraction 91/21 to lowest terms and gets the answer 4, people are not prone to regard him as having confused 4 with 13/3. Nor, if presentation of the case focuses on how he made the error, say he divided 7 into 91 and got 12 instead of 13, is anyone prone to say he confused the number 12 with 13. Neither is this linguistic incompetence. It is a (secondary) 'A'-mishap. But if the fractions are large (although not *that* large), intuitions waver in the same way they did in the Marsha case.

Real numbers are a mixed case. Suppose I calculate an integral and get the result

$$\int_0^{2\pi} \frac{d\theta}{5 + 4\sin\theta} = 2\pi$$

Intuitively, again, this is not a matter of confusing 2π with $2\pi/3$ (nor will it be seen as any other sort of A-mishap, if the example is developed enough to expose the exact source of my error, as we did with Jane's integrations). On the other hand, if I am calculating the decimal expansion of π and miscalculate in such a way that I mistakenly think π is a repeating decimal (after, say, the 400th place), some will say I *have* confused π with a particular rational number.

What about reference to mathematical individuals belonging to other categories, such as functions ($f(x) = 2x$) or sets ($\emptyset$)? Here it seems that errors are often classified initially as either primary A-mishaps or as linguistic incompetence. Suppose Jane thinks that $\int dx = x^2 + C$. Intuitively, this is usually taken to be a situation where Jane is confusing one class of functions with another. In particular, she thinks the first class of functions is the same as the second one (and she is wrong). Again, consider a particular set, $\mathcal{PPP}\emptyset \cap \mathcal{PP}\emptyset$, and suppose Jane thinks it is $\emptyset$. It is a common initial reaction to think Jane has confused one set with another. On the other hand, if Jane treats $f(x) = x^2$ as if it were $f(x) = 2x$, this is seen as linguistic incompetence (with respect to functions); the error, presumably, is too obvious.

Now, although A-infallibility *seems* to lapse with terms referring to large whole numbers, large rationals, real numbers under certain circumstances, functions, and so on, something strange is still going on. For despite the fact that when the cases sketched here are first presented to people, some have the intuitions I've described, it is usually possible to get them to concede that the error in question is actually (one or another) 'A'-mishap rather than a primary A-mishap.

For example, in the Marsha case, if one points out that because Marsha has miscounted zeros, and strictly speaking this is a notational matter (after all, the *number* 3 trillion doesn't have zeros in it), the person will withdraw the claim that a primary A-mishap has been made.

Similarly, although this is harder with anyone who is mathematically naive, one can usually get people to concede that the errors with functions or sets are actually errors with notation, and again the claim that a primary A-mishap has been made will be withdrawn.

So in many cases, one finds that with a little prompting, it is possible to elicit intuitions that support the extension of the apparent capacity for A-infallibility to large numbers, functions, and other mathematical objects. On the other hand, empirical terms are different: No prompting of whatever sort can achieve this shift in intuition when cases with empirical terms such as the ones given at the beginning of this section are presented.

But it isn't *always* possible to nudge a person away from the claim that in the mathematical cases we have been talking about a primary A-mishap has been made, and when it isn't, it is impossible to get him or her to recognize mathematical examples that crisply distinguish between 'A'-mishaps and primary A-mishaps, as we find it easy to do with empirical singular terms. This is because such people are invariably willing to describe the situations simultaneously as ones in which A- *and* 'A'-mishaps are occurring. Indeed, some go so far as to suggest that there is no difference between such mishaps in the mathematical case because the mathematical objects and the notation describing them are the same.[56]

Consider the Marsha case again, and consider someone who has just attributed a primary A-mishap to her. When it is pointed out that she made the mistake she made by miscounting zero numerals, such a person may claim that she committed an 'A'-mishap *too* (or that there is no difference between the kinds of mishaps in this case). And the same thing can happen when it is pointed out that the confusions involving integration or set theory are due to the mismanipulation of symbols.

So, with probing, people either find primary A-mishaps impossible in these cases or they find that the distinction between A- and 'A'-mishaps collapses. I should point out, though, that everyone seems willing to distinguish such mishaps (however described) from the

56 See Shapiro 1989, p. 171, note 38.

crude cases of out-and-out linguistic incompetence in respect to mathematical terms.

What is behind these intuitions? My hypothesis, which is one that will have occurred to many philosophers by now, is that use/mention "errors"[57] in the sense of Quine (1940, pp. 23–6) are at work here. But these errors, I think, are not merely a matter of confusing a term with what it refers to. Something a little more systematic and interesting is afoot.

Let us take the whole numbers first. There is a canonical notation for referring to such numbers, namely, the numeral system, and I think there is a preconscious tendency to identify numbers with such numerals – although most speakers will disallow the identification if pressed on it.[58] That is, the use/mention conflation is not between *any* piece of notation that refers to a whole number and that number but only between the canonical term referring to that number and the number. Thus, if someone (a child, say) asks what 116×7 is, the answer that will satisfy her is one in canonical notation (namely, '812'). I think this can explain the intuitions we have seen.

If numbers just *are* numerals in canonical notation, then the distinction between A- and 'A'-mishaps collapses in the case of numerals. Thus the hypothesis of the (preconscious) identification of numerals with numbers on the part of these individuals would explain why the distinction is not robust with respect to number terminology. But why is it so common to see errors with small numbers as evidence of linguistic incompetence, whereas, usually, errors with large ones are initially regarded as A-mishaps?

My conjecture is this: Mastering the manipulation of small numerals takes (relatively speaking) very little. And possession of that capacity is definitive of referential success. So when errors are made, they are seen as so crude that either they are (barely) 'A'-mishaps or (more likely) examples of linguistic incompetence. But errors with large numerals are easier to make, and where objects are (preconsciously) functioning autonominally, the role of the object as referent takes precedence over its role as referee – thus the tendency to take the error (initially) as an A-mishap.

Why does the role of the object as referent take precedence over its role as referee? My suspicion is that the use of a term on someone's part is intuitively taken to be prima facie evidence that he or she *knows*

---

57 I use shudder quotes to acknowledge that a number of philosophers will think that these are not errors at all but perhaps a just appreciation of what is actually going on. See Benacerraf 1965, Shapiro 1989, Chihara 1990, among others. Acknowledgments made, I drop the shudders hereafter.

58 The mark that a piece of notation is being treated as canonical (in a context) is that we tend to accept an answer in terms of it as final.

how to use it. That is, when something goes wrong with reference, our strategy is to look to the object first and turn to the term only when the problem can't be located in a mistake with the object. Notice that the way to dislodge A-mishap intuitions in the case of large numerals is to develop the example fully enough for it to become clear that the mishap is due to a mistake with the notation.

The intuitions regarding noncanonical numerical notation must be explained in terms of implicit (preconscious) assumptions about canonical notation. If someone has no problem manipulating numbers in a canonical notation, then we regard their access to numbers as unproblematical. Thus prima facie, the problem is a 'A'-misattribution via the piece of noncanonical notation – prima facie because if the story goes on to show that the person has trouble manipulating numbers in canonical notation, intuitions shift.

I think intuitions about rationals and real numbers should be explained in a similar fashion. First, trouble manipulating the notation can arise sooner with such numbers than it can with whole numbers. Thus there is a tendency to recognize the possibility of A-mishaps in a broader number of cases. Second, with rationals, the ideal choice for canonical notation might seem to be the fraction in lowest terms, but intuitively, fractions not in lowest terms (but containing simple numerals in the denominator and the numerator) are treated as canonical too. I think this is because the references of fractions are taken to be derivative from the references of the numerals in the numerators and denominators of such fractions.

Intuitions are less systematic in the case of real numbers, because there is no canonical notation for reals. The representation of reals in terms of decimal expansions is rarely treated as canonical, simply because such expansions are not notationally available in the way that the numeral system or fractions are (most of them – in a well-defined sense of "most" – cannot be characterized in a finitary way). Finally, intuitions about decimal expansions are often similar to intuitions about large numbers in numeral notation.

For example, my integral example with 'π' shows that 'π' is sometimes treated the way canonical notation for the numerals is. On the other hand, it is not unnatural for someone to ask what π *is*, and the answer, the ratio of the circumference of a circle to its diameter, or a nod toward the decimal expansion, will sometimes satisfy the questioner. This shows how intuitions will shift on some primitive constant names for reals (e.g., 'π'), sometimes treating them as canonical and sometimes not.

Here's an illustration of the claim I made about decimal expansions. If I manipulate an integral and get the answer .2373789443 . . . instead of .2374789443 . . . , most will regard it as not confusing the two numbers. But if I make errors in copying out the decimal expansions of these numbers, some will.

Notice that intuitions about functional ways of marking out reals, such as '$\sqrt{2}$' or

$$\int_1^2 \sqrt{x}\, dx$$

are treated (roughly) as canonical to a degree inversely proportional to the perceived complexity of the notation.

When it comes to sets and functions, things get worse. First, as with the reals, there really is no (intuitively acceptable) canonical notation. But worse, function notation and set-theoretic notation often wear important properties of what they refer to on their sleeves. Function notation often represents exactly how the functions depicted operate on their domains (think of polynomial notation), and set-theoretic notation often depicts how the sets depicted are built up from their constituents. Furthermore, there are often intuitions at work that treat functions as operations. This makes it easy to think of $f(x) = (x^2 + 1) - 1$ and $g(x) = x^2$ as *distinct* functions, and this in turn helps one think of functions as notational objects. Something similar can happen with set-theoretic notation if one has a tendency to think of sets as constructions. As a result of all of this, there tends to be a really active use/mention conflation at work in intuitions about sets and functions and, consequently, a complete collapse of the intuitions supporting a distinction between 'A'- and A-mishaps.

If my suggestions about the sources of these intuitions are correct, such intuitions should be sensitive to mathematical education. In particular, (former) child prodigies should not share all of them with the general population. Also, such intuitions should be affected by shifts in notational practices (e.g., intuitions about errors with scientific notation should differ from intuitions about errors with numeral notation even when the same *numbers* are involved).

If my analysis is only correct in its general outlines, it still poses puzzles for certain philosophers of mathematics. Let's assume we've shown that, apart from use/mention confusions, intuitions about mathematical practice support the assumption that the capacity for infallible A-attribution holds in the case of mathematical singular terms. This capacity is in some tension with the popular puzzle of referential access discussed in Section 2. For that puzzle was concerned with the question of how mathematicians succeed in talking about the same things. And quite a bit of philosophical work goes into trying to provide an explanation for this, as we saw. One invokes, perhaps, intellectual intuition. Or, perhaps, mathematical practice takes place in a language that enables us to uniquely pick out the objects we refer to.

But all this seems to be attempts to provide protection against an eventuality that *can't occur*. And by this I don't mean that in some way we can see that the mechanisms we use to refer to mathematical objects are mechanisms that can't fail. Rather, for singular terms at least, there don't seem to be any mechanisms for reference *at all;* and, perhaps consequently, there seems to be nothing in mathematical practice, as there is with empirical practices, to deal with potential primary A-mishaps. Notice the different ways in which the confusion about James and John would be corrected if the confusion were based on a primary 'A'-mishaps as opposed to a primary A-mishap. In the first case, we would explain that *his* name really is 'John'. In the second case, we would explain that *he* really is John. But this distinction in how errors with singular terms should be corrected seems *conceptually* alien to the practice of mathematics.

I should add that if we could have found our way to explaining the references of mathematical singular terms as being fixed by axiom systems, A-infallibility would be unsurprising. For terms whose reference is fixed *and accessed* only through descriptions do not submit themselves to primary A-mishaps. What makes A-infallibility puzzling in this case, if we have it, is that reference fixing through descriptions (axioms) seems to have been ruled out by the considerations raised in Section 4.

Furthermore, A-infallibility aside, it is puzzling that there should be such use/mention errors as we have divined in mathematical practice, at least on certain platonic views or on views such as Maddy's (1990). Use/mention errors are *errors* from such perspectives; therefore one's epistemic story must explain them, and the explanation must take the form of claiming that objects of one sort are *similar* to objects of the sort they are being confused with, or one must tell an epistemic story about how we get access to certain objects that explains why we might confuse them with other objects. One can thus understand *linguistic* use/mention errors, such as the one Quine attributes to Lewis – confusing implication between sentences with a sentential operator. One can also understand use/mention errors when they arise in contexts where one is talking about linguistic objects and using such objects at the same time.

But suppose the platonist thinks that our access to mathematical objects is through a kind of intellectual intuition. Then it is hard to see why they should be confused with the notation one uses to refer to them. Perhaps this sort of platonist could borrow a page from Quine and say: "Mathematicians attain precision because of the abstractness of their objects, and they confuse sign and object for the same reason. Physical things are palpably unlike their names; numbers and other mathematical objects, however, are not even palpable"

(1972, p. 50). Nicely put, but in fact, *values* are equally ineffable, and yet there are no use/mention errors operating in that domain. (None, anyway, that I can think of).[59]

These errors should be equally puzzling to someone like Maddy (1990) who claims that sets and other mathematical objects actually *are* the sorts of things one can perceive while denying that what is perceived is something linguistic. These intuitions suggest this is *not* how mathematical objects are perceived, for it seems both in such cases that primary A-mishaps would be possible and that the sorts of use/mention errors mentioned would not arise. Maddy's epistemic story makes these phenomena entirely inexplicable and it is precisely *that* which an epistemic story should not do. She *can* offer her suggestion as a reconstruction or a revision; but without an explanation for such intuitions, she cannot simply claim that this is what, sets, say, actually are. Maddy takes numbers to be properties of sets, and the same objections can be made in that case, as such properties are supposed to be perceived just as the sets are.

Kitcher's (1984) epistemic story fares slightly better. Although he, too, wants to use perception as a primary epistemic source of mathematical insight, he allows that mathematical knowledge is largely transmitted from generation to generation. Such transmission, taking place as it does in a written and oral medium, might provide an explanation for the use/mention errors that we've seen. The intuitions about 'A'- and A-mishaps, however, are still puzzling on Kitcher's story.

One result of these intuitions may be to tempt philosophers to consider the identification of mathematical objects in some way *with* the notation supposedly used to pick them out. In this case, such pervasive use/mention errors cease being errors at all. Furthermore, on this view, the intuitions about A- and 'A'-mishaps would make sense too. One well-known problem with this move, however, is the fact that mathematical objects, classically construed, far outstrip in extent our capacity to generate notation (although there are ways to get around this, as one finds in Chihara 1990).[60] Mere nominalism, on the other hand, when it is taken to be the simple *denial* that there are mathematical objects (as opposed to the *identification* of them with *some* of the notation supposedly referring to them), does not seem to do much by way of explaining such errors, for the same reason that the platonist's evocation of the impalpability of mathematical objects fails

---

59 Perhaps one detects a use/mention error in the practice of calling a note both the piece of notation and the sound produced. See Bay 1990, p. 5. But I know of no musical intuitions that turn on confusions of musical notation with the music itself, as we've found in the mathematical case.

60 See the end of Part II, Section 8, where I give reasons for not taking this approach.

to explain such errors. But perhaps the nominalist can point to our tendency to assimilate nonexistent objects to the ideas of such objects as a use/mention error similar to those discussed here.[61]

§6 Comparing Mathematical Terms and Empirical Terms II

Matters should change drastically when we turn to kind terms in mathematics. If numbers are being (preconsciously) identified with numerals, functions with the expressions for functions, sets with set notation, and so on, this won't imply conflations of kind terms with kinds – that is, 'number' with the set of numbers, 'group' with the class groups, 'real function' with the set of real functions, and so on. And, indeed, no such use/mention errors seem to be at work.

Further, if we take the notions of A-mishaps and 'A'-mishaps as generalized to empirical-kind terms and then apply them to mathematical-kind terms, they ought to survive nicely. For even if the subject matter under discussion is (preconsciously) *notation*, there is still a distinction to be made between mistaking one piece of notation for another and misapplying a term to one piece of notation when it should really apply to another. And this is exactly what we find: The distinction between 'A'-mishaps and A-mishaps is intuitively robust with regard to mathematical kinds.

On the other hand, although the survival of this distinction seems to make kind terms in mathematics akin to empirical-kind terms, we will find disanalogies. Among other things, we will find that the potential for referential dysfunction is far greater with empirical kinds than it is with mathematical kinds.

Let me show how the various sorts of A- and 'A'-mishaps are possible with mathematical kind terms. First, 'A'-mishaps: I may confuse the definition of a ring with a group. "Groups," I may say in my muddled way, "have two operations. . . ." Examples of this sort can be multiplied ad infinitum.

Next, A-mishaps: By virtue of a complex and subtle proof, I may believe that the fundamental group of a space B is a certain abelien group A. And I may be wrong. Similarly, on the basis of a subtle and complex modification of Gelfond's proof, I may convince myself that π^{π} is algebraic.[62] And I may be wrong. These examples, and others like them, do seem to be examples of A-mishaps. And it is very hard

---

61 For an illustration of this, see Quine 1951b, p. 2.
62 Notice that these sorts of examples allow secondary A-mishaps with singular terms. We *can* be mistaken about what properties a particular mathematical object has, although, it seems, we can't confuse one mathematical object with another. If mathematical objects *are* being preconsciously identified with notation, it is clear how this otherwise strange state of affairs is possible.

to dislodge this impression, even when the details of the sources of the errors are given.

For example, consider the discussion of Cauchy's mistakes in the theory of the convergence of Fourier series in Grattan-Guinness 1970. One mistake among several stands out: Cauchy systematically failed to distinguish between series of constant terms and series of functions. But it certainly isn't that Cauchy *confused* functions with constants. Rather, it is simply that he overlooked certain additional complications involved with functions. There is no way here to nudge intuitions so that Cauchy's errors can be seen as examples of 'A'-mishaps. So when he concluded that the Fourier series of any discontinuous function is not convergent to it, this must simply be seen as an A-mishap. Cauchy thinks that every function that is Fourier-representable is continuous, and this is false.

Thus far, kind terms in mathematics seem on a par with natural-kind terms in the empirical sciences in a way that mathematical singular terms and empirical singular terms aren't. But now we turn to the differences, and they are substantial.

First, there seem to be a priori limits on the A-mishaps possible with mathematical kinds, and this seems linked to a much deeper fact about mathematical kinds – to wit, that their extent is in crucial ways not discovered but *stipulated*.

One can be wrong about whether a group has this property or that property; one can be wrong about whether this class of functions is identical with that class of functions. We can even imagine (apart from questions of mathematical necessity) that one could have been wrong about whether $\sqrt{2}$ was rational. But one cannot discover that numbers are not sets (or that they are). One cannot discover that certain functions are a kind of real number. One cannot discover that certain mathematical objects are planets. Nor did anyone discover that Euclid's line is our contemporary real line!

What *can* be discovered is that certain identifications of mathematical kinds (apart from an arbitrary element) are available. That is, once it is discovered that certain sorts of identifications are possible, what can follow are acts of *fiat*. One can *identify* the set of numbers with (one or another) set of sets. One can identify real numbers with certain functions. One can identify mathematical objects with certain physical ones (if there are enough of them).

The mark of what I am calling "fiat" here is twofold. First, there is usually ineliminably arbitrary choice available in how to make the identification. Second, the identification is not forced on us by proofs.

This ubiquitous mathematical phenomenon seems to have been first brought to the attention of the philosophical community by Benacerraf (1965). The case he noticed and publicized was that there are many ways to embed numbers in set theory. But the same holds of the embedding of geometry in real number theory afforded by analytic geometry. If we try to *reduce* three-dimensional Euclidean space to triples of real numbers, we find that there are infinitely many ways to do it (notice that infinitely many coordinate systems are available). The issue is identical in spirit to the one Benacerraf discussed. Indeed, there are many such cases.[63]

Contrast this situation with that of natural kinds in the empirical sciences. First, there doesn't seem to be any limit on the sorts of mishaps possible. We *could* discover that lightning is actually a certain kind of (fast-growing) fungus. Quite a lot of our picture of the world would have to be wrong for this to happen, although it is certainly not impossible. But nothing similar seems available with kind terms in mathematics. Rather, as I have said, beyond a certain point one does not discover – one stipulates. We could not *discover* that irrational numbers are actually Hilbert spaces. (It's impossible for us to stipulate this by the way – the objects are too unalike in properties – but that isn't at issue here.)

There is another respect in which mathematical-kind terms differ from empirical-kind terms. It seems that the referential scope of mathematical-kind terms at a time is sometimes taken to be restricted, to some extent, by the metamathematical views of mathematicians using the terms at that time. For example, historians of mathematics do not seem to assume that the term 'function' referred in the hands of d'Alembert or Euler to the same things it refers to now.

This issue is worth elaborating on. We find the following passage in Langer 1947, p. 4: "While . . . the function and the analytic formula were one to d'Alembert, the function was thought of as a graph by Euler, and probably meant something else again to still another." Again, we find in Grattan-Guinness 1970, p. 4: "Now the calculus with which they [Bernoulli et al.] worked was essentially a *calculus of operators on algebraic expressions;* the operators were those of differentiation and integration, whose inverse character had been the great discovery of both Newton and Leibniz." Similar remarks hold about the notion of *number,* the notion of *set,* and numerous other mathematical notions.

To make clear how different the situation here is from the case with empirical-kind terms, let us distinguish the talk of notions and ideas – what I call the intensional talk – from the actual objects this talk picks out.[64] Clearly, such ideas, with respect to many terms, and

---

63 Recall the second paragraph of note 46.
64 Of course, some historians of mathematics identify these things. See Kline 1972, where we find: "One of the great Greek contributions to the very concept of mathematics was the conscious recognition and emphasis of the fact that mathematical

with them definitions and "truths," change over time. But as we've learned from work on the reference of empirical-kind terms, from this on its own it doesn't follow that what is referred to by such terms must change. Consider, 'gold', 'water', and so on. There are strong intuitions that such terms pick out what they pick out despite variation in "intension."[65]

By contrast, mathematical-kind terms seem able to shift in their extensions: It seems clear that the extension of 'function' is taken to have undergone a steady expansion during the eighteenth and nineteenth centuries. It simply won't do to suggest that what 'function' picked out for Euler was the same as what it picked out for Riemann. No one talks this way. 'Gold', on the other hand, is very different. Although the tools the Greeks had for picking out instances of gold were far more restricted than the tools we have – for example, we can recognize the presence of gold in seawater and they couldn't – and even though gold in some of its forms (e.g., as a gas) would have been conceptually alien to them, nevertheless, their term currently translated as 'gold' is taken to pick out the same stuff as our term does.

Now, although we allow the practices of the mathematician to fix the extent of what his or her kind terms refer to, as we've seen with the example 'function', there are limits, because the particular objects the terms pick out survive reworkings of those terms. Even if we take d'Alembert's concept of function to be as Langer suggests, still, it includes $f(x) = x^2$, the same function we deal with today. In some sense, the referential power of singular terms in mathematics is more fundamental than the referential power of kind terms.

There is a certain tension, therefore, between the intuitions here and those discussed in Section 5. There I pointed out that historians generally regard reference to a purported mathematical object, or even a class of such objects, to be successful right from the introduction of terminology that is taken to pick such objects out. I gave examples ranging from (subclasses of) the natural numbers in the hands of primitive tribes to the use (and misuse) of complex numbers.

But now I have noted that historians are equally prone to recognize that mathematical notions can shift over time, and with this shifting can come shifts in extension. Langer would not assume, I imagine,

entities, numbers, and geometrical figures are abstractions, ideas entertained by the mind and sharply distinguished from physical objects or pictures" (p. 29). For similar sentiments, see my quote from Bos 1984 in note 81. It is relatively clear (to philosophers, anyway) that such an identification is problematic.

65 See Kripke 1980 and Putnam 1975b, 1975c. I should point out that this picture won't *quite* do for empirical terms without some tinkering, which I avoid doing now. Still, the distinction – between how such terms seem to operate referentially and how mathematical-kind terms seem to operate – remains.

that d'Alembert's notion of function contains in its extension an everywhere continuous, nowhere differentiable function (for no analytic formula corresponds to such an item). Despite this, and despite the fact that such ontologically different sorts of things have rather different properties (e.g., the number of real functions in the contemporary sense is far greater than the number of analytic formulas), no one seems to have any problems identifying the objects studied by d'Alembert with the objects we currently study or identifying the theorems d'Alembert showed with the theorems we currently use; indeed, it is worth stressing that such identifications do not usually lead to trouble even among historians of mathematics.[66]

Let me sum up what we've covered so far. Mathematical-kind terms are a little more like natural-kind terms in one respect than mathematical singular terms are like empirical singular terms: Primary A-mishaps are intuitively possible. On the other hand, mathematical-kind terms seem much more restricted referentially by mathematical practice (perhaps in Kitcher's sense) than seems possible with natural-kind terms. And how wrong we can be about mathematical objects seems to be limited by the fact that the extensions of mathematical-kind terms are hemmed in by what look like a priori constraints and stipulations. Finally, some mathematical-kind terms seem to have the peculiar property of being able to shift radically in their extensions, going from having objects of one sort there, say, formulas, to objects of another sort with rather different properties there, say, subsets of $\mathbb{R} \times \mathbb{R}$, while simultaneously (pretty much) still containing in their extensions what was previously there (e.g., $f(x) = x^2$); as a consequence, the results established in terms of kind terms with their previous extensions are taken to be the same as those versions of these results in terms of the kind terms with their new extensions.

There is one last way in which mathematical-kind terms seem to differ from their empirical brethren. Recall that I have discussed, broadly speaking, three kinds of referential mishaps – A-mishaps, 'A'-mishaps, and out-and-out linguistic incompetence – and I have attempted to use such possibilities to compare mathematical and empirical nomenclature. But the history of science reveals another possibility. We can discover that our attributions fail, not because we have picked out the wrong thing, but because the theory surrounding such a term is faulty; we are entirely off in a widespread and global fashion in our descriptions of what is going on there. Then something like referential legislation seems to go on *in empirical science:* We find

---

66 Contrast the case with studies in empirical science where the dangers of anachronism are ever present. One might suspect that the history of mathematics still awaits its Kuhn, but I think this would be a mistake. Something different is going on.

ourselves dropping terms altogether, or seriously altering our claims about what the objects under study do, and when we are in touch with them. And of course, we may alter our claims about what kinds of objects there *are*. There is epistemic failure here too, but of such a radical kind that it is simply inaccurate to describe the situation as one in which we have mistakenly called something that was not an A an A. That description turned on presuming the reference of the term fixed despite our epistemic failure, and this doesn't seem to be a reasonable assumption here. Also, intuitively, what has happened in these sorts of cases doesn't involve linguistic incompetence *either*. Rather, it is that the terms we were using did not pick out anything "real."

What do I mean by "real" here? I'll say something fairly crude now, with the intention of refining the remarks later, in Part II of this book. Theoretical objects in the sciences are ones that we are *instrumentally* connected to. Regardless of how remote their operations are from our senses, we usually causally interact with them, and usually by designing instruments that are causally sensitive to their machinations.[67] Now, when a radical reevaluation of the sort described earlier takes place, one thing that happens is that we have to *redescribe* what it is our instruments have been detecting. For example, once we discarded the notion of caloric fluid, we had to redescribe what it was we were actually recognizing when we thought we were recognizing the presence of caloric fluid.

And this is always the case. To drop an empirical scientific term that is even slightly entrenched is to require a redescription in *epistemology*, a redescription of what the instruments associated with that term by our scientific practice are doing. And, of course, the more entrenched a term is, the more involved the needed redescription becomes.[68] A requirement of any term that is real in this sense is that

---

67 'Instrument' here is to be understood broadly. I not only have in mind devices that measure something the way that we can measure mean kinetic energy, mass, and so on, but also the wide array of tools and devices that rely on theoretical objects as instruments to manipulate something else, for example, as when we manipulate atoms by firing photons into them. I mean, in short, our entire apparatus for causally interacting with the world about us.

68 Of course this isn't always possible. We may find that we must drop a term and the theory surrounding it for certain reasons, and accept that certain phenomena, even phenomena involving our own instruments, have lost an explanation they once had. And there may be no substitute on the horizon. We could decide, by the way, that certain instruments pick out nothing at all, or that several sorts of thing were involved. How the explanation forthcoming treats the instrumental practices to be explained is quite open.

Also, it seems an interesting sort of schizophrenia is possible. Arguably, the Copenhagen interpretation involves a radical (quantum) shift in scientific nomenclature for atomic phenomena while keeping the classical nomenclature for the scientific instruments causally interacting with that phenomena.

it (eventually) play a role in our explanation of how our instruments work.

Despite Lakatos 1976a, this sort of misattribution *seems impossible* in the case of mathematics, and the Frege example cited in Section 4 illustrates this. Once one has terminology to refer to certain mathematical objects, it is not possible for theorem proving to go in such a direction that mathematicians realize that a serious terminological overhaul is necessary – that all the terms they are using in a particular branch of mathematics perhaps pick nothing out at all, or that all the definitions in that branch used to pick out things fail to pick them out.

This is not to deny that mathematicians can recognize that definitions are inconsistent, or that the notions they are using are not the most appropriate ones for studying a domain, as we've seen. It does not even rule out the replacement of one term by a homophonic duplicate with a different extension. Nor does it rule out the possibility that an entire branch of mathematics can fall into oblivion. But none of these possibilities resembles what happens in the empirical sciences: the wholesale dumping of a large body of theory. Furthermore, subtle researches in mathematics *never* involve the technology of access that we find in empirical science, nor is there a requirement of rewriting the explanation for this technology when theory changes.

We have found the following: The use of mathematical terms to pick out particular mathematical objects seems to be embroiled in use/mention errors. Furthermore, if we factor out the use/mention errors, we find that our use of singular terms gives us A-infallibility. Kind terms don't quite give us A-infallibility, but what they do offer are curiously restricted referential scopes, in two respects. First, the range of such terms seems more restricted by mathematical practice than by what exists mathematically – in sharp contrast with empirical-kind terms; and second, the possibilities for various sorts of mishaps seems restricted in ways not available for empirical terms.

These results, and the ones discussed earlier in Section 4, seem to exclude two possible ways of explaining how mathematical terms refer. The first possibility apparently excluded is to take the references of mathematical terms to be fixed by axiom systems or to be fixed by something slightly broader, such as practices in Kitcher's sense. The second possibility apparently excluded is to take the references of mathematical terms to be fixed by some sort of (quasi-causal) mechanism that connects such terms to the objects they refer to.

A methodological caveat: Of course, the intuitions and practices I used to argue for these results are raw. That is to say, any semantic theory of mathematical discourse is allowed to explain them away using whatever resources are available to it – and, in doing so, it may reclassify the apparent category the intuitions and practices are in.

Here's an example from linguistics of the sort of thing I have in mind: What *seem* to be grammatical intuitions may be dismissed as due to performance factors such as memory limitations. Similarly, what seem here to be intuitions that bear on the mechanisms of reference at work with mathematical terms may be explained away in some other way entirely. This is certainly the sort of strategy that a nominalist, for example, would want to pursue, unless, of course, such a nominalist was offering nominalism as an explicit *reconstruction* of traditional mathematical practice. However, my motive in describing these intuitions and practices is to point out how, prima facie, they exclude certain otherwise natural explanations of how mathematical terms operate.

§7 The Epistemic Role Puzzle[69]

We have discussed the puzzle of referential access and followed it up with a number of puzzling intuitions about reference that, for the most part, have not been noticed by philosophers of mathematics. As it turns out, the situation is analogous when it comes to the epistemology of mathematics, though somewhat neater. There is a pair of puzzles about knowledge, one pretty much unnoticed in the literature and the other regularly cited.

But I'm going to reverse the order of presentation and write about the generally unnoticed puzzle first, before turning to the one popularized by Benacerraf. The unnoticed puzzle is this: Given standard mathematical practice, there seems to be *no* epistemic role for mathematical objects. Although attention is rarely drawn to this puzzle, I think awareness of it at some level on the part of the philosophical community helps explain why the Quinean tendency to treat mathematical knowledge as entirely inferential from the instrumental needs of empirical science seems a reasonable tendency to have.

The second puzzle, deeply influential in the philosophy of mathematics, is this: On certain philosophically plausible views of what knowledge is, metaphysically inert objects seem to be objects we *cannot* know anything about. Consequently, if mathematical objects are metaphysically inert, we cannot know anything about them. The unremarked puzzle first.

Proving theorems is not easy. That is to say, establishing mathematical truths when they go beyond simple computational facts is never a

---

69 George Boolos has drawn my attention to the family resemblance between my epistemic role puzzle and Wittgenstein's beetle in the box example. See section 293 in Wittgenstein 1968.

trivial matter. And yet, somehow, as I observed in Section 1, there seems to be no role for mathematical objects in this process.[70] It is not merely that mathematical objects do not seem causally involved in the processes we use to learn about their properties; it is that they seem to play no role at all. (Imagine that mathematical objects ceased to exist sometime in 1968. Mathematical work went on as usual. Why wouldn't it?)[71]

An important indication of the absence of an epistemic role for mathematical objects is the tradition, present since antiquity, of regarding diagrams, pictures, and mathematical notation generally as devices that simultaneously operate referentially and as *heuristics*.

Consider a couple of examples. A diagram of a triangle contains parts (the lines, the angles) that represent other mathematical objects that bear significant mathematical relations to the triangle.[72] Certain points are represented, such as the vertices of the angles making up the triangle, and other points may also be marked out (such as midpoints, for example) by dots or line segments.

For another example, consider the following integral

$$\int_1^2 x \, dx.$$

This actually refers to a particular real number. But certain mathematical objects other than the one the symbol as a whole picks out are semantically indicated by the notation. I refer, of course, to the num-

70 I forgo (but only for the moment) the suggestion that might be made that *notational* objects, such as numerals, computer representations, and so on, are the mathematical objects needed to play the epistemic role under discussion.
71 Thanks are due to George Boolos for making me aware of a forthcoming article by Edward Nelson in which the point I make here is made with practically the same example. See Nelson in press.
72 In the geometrical case we can be tempted to regard the lines and points depicted as "parts" of the triangle just as the drawings of the lines and points are parts of the drawing of the triangle. And I have to say that even a cursory glance at mathematical notation makes one start to think along the lines found in Wittgenstein 1961. Nevertheless, any suggestion that the spatial location of the semantically significant parts of mathematical notation pictures the part–whole relation among the mathematical objects so pictured is hopeless. In geometry, for example, which mathematical objects (points, lines, geometrical figures) are taken as primitive and which as constructs is rather arbitrary. The suggestion looks even worse when one considers other mathematical notation such as integral representations of particular real numbers – for example

$$\int_1^2 dx.$$

I find in Kitcher's (1984, p. 130) musings on notation something kindred to my own thinking. However, he wants to suggest that "not only do sentences which occur in mathematics books describe ideal mathematical operations (more exactly, the ideal operations of an ideal subject) but, in producing those sentences, the mathematician may be engaged in *performing* those operations." This sounds like an idealist version of the picture theory of meaning. It won't work.

bers 1 and 2, as well as the particular function $y = x$. The "diagram chasing" that goes on in category theory is yet another notable example of this rather ubiquitous sort of thing.

Such mathematical diagrams and terminology are striking in that, to a large extent, what grammar they have corresponds to mathematically significant relations. Indeed, informal mathematical notation is a prototype of the syntactic representation of logical form that one finds in formalized systems. Consequently, manipulating such notation can yield mathematical insights.[73]

We have, therefore, in mathematical notation, a terminologically rich avenue for recognizing truths about mathematical objects. But notice the following:

(1) Classically, the diagram or term is not taken to *be* a mathematical object. This point applies both to the particular tokens and types of diagrams and terms used.

(2) Mathematical objects themselves seem to play no epistemic role, but only a role here that is strictly semantic: The diagram has "parts" that refer to these objects.

(3) Insight gained through manipulating such notational objects does not involve any manipulation of, or comparison with, the mathematical objects themselves. There is no sense in which, or so it seems, the diagram, to the extent it functions as a model, is compared to the original it is a model of.

I have focused here on mathematical notation as it arises in informal practice. But the point is even more transparent when one turns to (interpreted) formalized systems. Theorem proving goes on *autonomously:* There seems to be no dependence on the objects the terms (occurring in the theorems) refer to. And when one looks at how the axioms or postulates are established, again, there seems to be no process that involves (epistemically speaking) the mathematical objects these axioms or postulates are about.

The point simply is this: Mathematicians have epistemic practices that justify their beliefs in the truths they establish. But nothing in these practices seems to involve mathematical objects as traditionally construed. All the work seems to be done by the contemplation and manipulation of *notational objects* that are *not* taken to be the objects under study.

If a philosopher (specifically, an epistemologist) pressures a mathematician to give an explanation of how he knows about the objects he talks about, and the mathematician does not want to invoke nonnatural intuitions, he may be cowed into going formalist or taking

---

73 See Frege 1953, p. iv^c.

such notational objects as self-referential, as themselves instances of the actual mathematical objects they are supposed to be referring to. But clearly this is not the natural attitude, and there is a prima facie good reason: The mathematician does not want mathematically extraneous properties that such objects have (being made of chalk dust, having visually indetectable curvature) attributed to *mathematical* objects.[74]

I claimed earlier that this puzzle helps explain the appeal of the Quinean attitude toward mathematical objects, and now we can see why. If mathematical objects play no role in the epistemology of mathematics, then this supports a prima facie case that the only epistemic role possible for them is one mediated via the theoretical needs of empirical science.

I have described this as a puzzle. Perhaps, though, it is not a puzzle *yet*, only a set of facts that may be explained in any number of ways. But certainly something here needs to be explained, namely, given that the traditional practice of mathematics is as I describe it, what exactly is it in traditional mathematical practice that mathematical objects do? I call this puzzle the *epistemic role puzzle*.

The puzzle just discussed and the set of referential puzzles given earlier together explain the source of the impression that mathematical objects are metaphysically inert. They also make acute the question: *What is the point, if any, of mathematical objects in classical mathematical practice?*

§8   Benacerraf's Puzzle

I now turn to the second epistemic puzzle. One of the most influential papers in the philosophy of mathematics is Benacerraf 1973. As Maddy (1990) points out, Benacerrafian considerations have motivated much of the work in philosophy of mathematics. Field 1980, Bonevac 1982, Gottlieb 1980, Hellman 1989, Kitcher 1984, Resnik 1981, 1982, Chihara 1990, and Maddy herself – all allude to such considerations to one degree or another.

The problem Benacerraf poses is pretty clear. We start with the assumption that our best current theory of how we know what we know is one on which "for X to know that S is true requires some causal relation to obtain between X and the referents of the names, predicates, and quantifiers of S" (Benacerraf 1973, p. 412). Add to that the picture given earlier of acausal mathematical objects, and it is easy to draw the conclusion that we cannot know anything about such objects.

Although Benacerraf's reliance on the causal theory of knowledge rather dates his exposition of the problem, the list two paragraphs up

---

74 Frege makes this point, see his 1953, p. vii[e].

shows that it doesn't seem to have dated the problem. Part of the reason for this is that when one turns to later epistemological views, the problem remains;[75] such theories generally require the object(s) of knowledge to play a role in the explanation of how we acquire that knowledge, and the kind of role I mean here can be described this way: "[I]n order to be dependable, the process by which I come to believe claims about xs must ultimately be responsive in some appropriate way to actual xs" (Maddy 1990, p. 44).

I hope it is clear that Benacerraf's puzzle is distinct from the epistemic role puzzle. Benacerraf's puzzle presupposes one view or another of what processes by which we gain knowledge must look like, which the epistemic role puzzle clearly does not. And, of course, this allows for a rather trenchant response to Benacerraf's puzzle: Because it is perfectly obvious that we *do* have mathematical knowledge, any epistemological theory that renders such knowledge implausible *must* be wrong.[76]

As a result, it may seem surprising that Benacerrafian worries have motivated so much of the recent literature in the philosophy of mathematics. For these worries seem to be a two-edged sword that may cut either the epistemic theory that rules out mathematical knowledge of metaphysically inert objects or any semantic view of mathematics that takes its terms to refer to metaphysically inert objects.[77] (This point seems to apply equally well to the naturalized theories of reference mentioned in Section 2: Perhaps metaphysically inert objects are inappropriate referents for mathematical terms, or perhaps such theories themselves are in trouble.)

We might try to argue that the apparent deadlock here is broken by the naturalistic constraints mentioned in Section 1: We are supposed to tell a naturalized story about how we know about and refer to such objects, and no such story is given by just claiming that mathematical knowledge is a species of knowledge. Mathematical knowledge, on such a view, "would be a complete mystery" (Chihara 1990, p. 193). If the traditional story of acausal, abstract objects outside of space and time prevents any naturalized story, then they have to go.

Historically, however, there *is* such a story available, and it is Quine's: Mathematical objects are, epistemologically speaking, posits. Consequently, there is no need to explain how the process by which

75 For example, a view such as that found in Dretske 1981 cannot on his admission (note 3, pp. 264–5) handle mathematical (and logical) knowledge.

76 Burgess 1983 runs this argument on pp. 100–1. One also senses the argument at work in Tait 1986, pp. 350–1. A response to this line may be found in Chihara 1990, pp. 192–3.

77 Indeed, as we've seen, different philosophers have swung this sword in different directions. See Maddy 1990, pp. 41–50 for further discussion on this. Also see Wagner 1982, p. 261, where the deadlock is explicitly noted.

we come to believe in mathematical objects is ultimately responsive in some appropriate way to actual mathematical objects.

But philosophers of mathematics seem not to have liked Quine's approach on this point, and the evidence of this is that so many are sensitive to the Benacerrafian dilemma and so few of those consider the Quinean response, even by way of indicating how it should be disabled.[78]

Rather, it turns out that most (recent) philosophers of mathematics have swung the sword at the traditional view of mathematical objects. Maddy (1991) sees a consensus emerging: "[S]ome form of ontological tinkering can defuse Benacerraf's dilemma without sacrificing standard mathematics" (p. 156). Indeed, the list I cite of philosophers influenced by Benacerraf 1973 is simultaneously a list of philosophers who tinker in the fashion Maddy means.

I detect three strategies, which often overlap. First, one may replace the mathematical objects with *ersätze* that are epistemically available to us. I put the recent structuralists and modalists in this category, as well as Chihara 1990. Kitcher 1984, intuitionists, and certain time-slices of Putnam also belong here.[79] Next, one may remove the offensive objects from our sight altogether (while giving reasons, of course, for why we don't need them). Field 1980 exemplifies this nominalist strategy; in some sense so do modalists such as Hellman 1989. Substitutional evasions of ontology such as the one found in Gottlieb 1980 belong here too. Finally there is the strategy of trying to show that mathematical objects are *not* epistemically inaccessible, that in fact, they are (naturalistically) accessible by "broadly perceptual experience" (Maddy 1991, p. 156). Maddy explicitly belongs here, and on her interpretation many other philosophers of mathematics, such as Field 1980 and Chihara 1990, belong here too.[80] These strategies are all ontologically radical in the sense given in Section 2.

§9 Comparing Puzzles

We now have two sets of puzzles. Benacerraf's puzzle, which is a puzzle about epistemic access, and the puzzle of referential access, which

---

78 Maddy (1990, p. 45) notes this, and attempts to explain the apparent oversight. I will discuss Quine's approach and its drawbacks shortly.

79 I am thinking of, for example, Putnam 1977; he writes: "Models are not lost noumenal waifs looking for someone to name them; they are constructions within our theory itself, and they have names from birth" (p. 25).

80 Maddy (1991) divides up the terrain differently. She claims that the broad consensus involves the replacement of the traditional mathematical object by something for which "the most basic knowledge is said to be gained by broadly perceptual experience" (p. 156). My own feeling is that many of the philosophers invoked will find her description of them somewhat procrustean.

I discuss in Section 1, is one set. On the other hand, there is the epistemic role puzzle and the various intuitions and practices regarding reference that seem to offer prima facie evidence against various positions for what the referential mechanism in mathematics could look like. Let us call the first set the *traditional puzzles* and the second set the *new puzzles*.

Now, as I have already intimated, these puzzle pairs are in tension with each other philosophically in that solutions to the traditional puzzles invariably fail to be solutions to the new puzzles; and, I will suggest, it is this that explains why so much work in the philosophy of mathematics has an unsatisfying flavor, despite its (often) high technical caliber.

I will draw the conclusion that focusing on the traditional puzzles rather than on the new puzzles has misled philosophers of mathematics for nearly two decades. Too much recent philosophy of mathematics has been an attempt to replace mathematical objects with something to which we have naturalized referential and epistemic access without showing that mathematical practice *needs* objects of this sort in the first place.

Nominalist programs are open to a similar objection. They too have often started from the premise that what is wrong with linguistic realism is that it postulates objects to which naturalized referential and epistemic access is impossible while assuming (apparently without argument) that mathematical practice requires that the mathematical objects so needed are ones to which naturalized referential and epistemic access is necessary.

Let us start, then, with a direct comparison of the two puzzle sets. The referential-access problem motivates solutions that enable us to pick out such objects uniquely. On the other hand, the epistemic-access problem motivates solutions that replace mathematical objects with objects that we do have epistemic access to. Of course, these puzzles can also be solved in a more radical way: nominalism.

The new puzzles don't support the generating of philosophical programs so much as bewilderment about what is going on in the practice of classical mathematics; they also engender puzzlement about the cogency of the philosophical solutions just mentioned, for consider:

(1) If mathematical objects have no epistemic role, why do we need them in mathematical practice at all? Why can't we just continue to use the theorems just as classical mathematicians do, but concede that the mathematical idioms are referentially empty, and draw attention to the lack of an epistemic role for mathematical objects in mathematical practice as proof of this fact?

Related to this is the following worry:

(2) What is the point of attempting to design languages that enable us to pick out such objects uniquely when they do nothing epistemically in mathematical practice? (What are we afraid of happening if we don't refer to the right ones? What could go wrong?)

Recall the approach discussed in Sections 2 and 3 that utilized lists of truths (in one or another language expressively richer than the first-order predicate calculus) to pick out the mathematical objects referred to. This descriptivist solution for the problem of reference seemed to concede our noninteraction with mathematical objects. But then what is the point of going though all this? After all, if no epistemic explanation for our access to mathematical truth that involves mathematical objects is available, then if any explanation for why we can trust the mathematics we produce will not include the mathematical objects the mathematical truths are about, why do we need to work so hard to make sure our mathematical terms uniquely pick them out? Again, why not just concede that referential idioms function emptily when it comes to mathematical objects?

(3) What is it about mathematics that enables so many of *its* theoretical posits to be ones to which misattribution is impossible? Doesn't this also show that there is something *empty* about our commitment to mathematical objects as mathematical practice currently stands? One can't imagine things getting much worse than they did for Frege – and yet he *still* succeeded in referring to sets. (So what are nominalists like Field so hard at work at? Aren't we already there?)

And related to this is this worry:

(4) Why should we have epistemic worries about mathematical objects of the sort Benacerraf has, if so much in the practice of mathematics suggests that A-mishaps in respect to mathematical objects are rarely possible, and when possible, severely restricted in scope? If so much is stipulated, what is there to be afraid of? Again, what can go wrong and how?

I claimed that philosophical solutions to the traditional puzzles are invariably not solutions to the new puzzles. Let me illustrate this point with a few examples. First let us take the nominalist solution. If the nominalist solution came in a form that showed that the mathematical commitment to mathematical objects really was trivial – that the apparent linguistic commitment to such objects witnessed by the appearance of certain terms in mathematical truths was easily eliminable – we could relax, reckon mathematical talk of numbers as similar to vernacular talk of 'sakes', and be done with the problem.

But what has come to be called Quine–Putnam indispensability thesis, coupled with the Quinean criterion for existential commitment, seems to show that nominalist solutions must run deeper. Like Field's approach (1980) or Gottlieb's (1980), they must call for a substantial rewriting of mathematical practice. The idioms cannot be taken at face value. But consequently, such approaches cannot explain what is so puzzling about mathematical practice in the first place – for example, that mathematical objects *already fail* to play an epistemic role in mathematical practice. They sidestep the issue by replacing that practice with something else: That such a project could be successfully carried out may not even be an indication of why traditional mathematical practice has this puzzling feel, for such a project, conceivably, could be carried out on an empirical subject where up until the time the new practice supplanted traditional practice, one *was* committed to the objects in question.

Similar remarks apply to Maddy 1990. She works hard to show that (some) sets are objects that one can perceive. To this purpose, a bit of psychology is employed, and perhaps one is convinced that, for fairly elementary mathematical objects, this is possible. But there is simply no indication what light is shed on mathematical practice by this move. It is hard to see what further role perceivable mathematical objects would (or do) play in mathematics beyond the role already conceded for heuristic notational devices.

Let me put the matter another way. Recall the requirement I quoted from Maddy 1990 in Section 8, namely, "[I]n order to be dependable, the process by which I come to believe claims about xs must ultimately be responsive in some appropriate way to actual xs." One does not show this in the case of mathematical objects *merely* by showing that (some) mathematical objects are objects that one can take to be perceivable. The reason is that, although perception is (generally) a method by which we gain knowledge, it doesn't follow that something that we know about an object we perceive is something we learned *through* perception. For example, we may know it already. What one needs here is a demonstration that reliability in mathematical practice, at least in the case of the mathematical objects Maddy presumes we can perceive, derives from perception. And this, to say the least, has *not* been established by any considerations raised in Maddy 1990.

Furthermore, when the matter is put this way, it looks implausible. Consider Greek geometric practice as compared to that of their Egyptian predecessors. Egyptian geometers apparently *did* rely on perception to derive their conclusions. This is why they got many of them *wrong*. Greek practice treated the perceptual items as heuristics, and so we have ever since.[81]

81 The claim that diagrams were heuristic for the Greeks is compatible with the fact that "constructions" in the sense of Bos 1984 were a significant mathematical study until at least 1750. See especially note 2, p. 332, where Bos writes, "It should be stressed that the operations involved in geometrical construction as here consid-

§10   Quine's Approach I

I have claimed that the new puzzles are not solved by *any* attempts at solving the old puzzles, and despite my nods in Quine's direction, this applies to his approach as well. But Quine's approach calls for extended treatment, for two reasons. First, as I mentioned, it solves Benacerraf's puzzle (before, in fact, the puzzle was posed in Benacerraf 1973), and so one needs to get clear why recent philosophers of mathematics see it as inadequate. Second, it comes *closer* than any other approach to handling the new puzzles, and consequently it is instructive to see how it fails and why. I discuss Quine's approach in this and the next section. My criticisms and observations come first, and then I turn to certain criticisms of Quine that have been made in the literature.

To start, we may ask what Quinean epistemology looks like, and for that we should turn to what he says about evidence. Here is a characteristic passage:

Having noted that man has no evidence for the existence of bodies beyond the fact that their assumption helps them organize experience, we . . . [do] . . . well, instead of disclaiming evidence for the existence of bodies, to conclude: such, then, at bottom, is what evidence is, both for ordinary bodies and for molecules. (Quine 1955, p. 251)

Quine has repeatedly claimed that evidence for the existence of mathematical objects is the same kind: It is successful at organizing experience. This seems to show that for Quine physical objects and abstract objects are on a par epistemologically. And this does seem to be his spirit: Experience, neural though it may be, is organized by *posits* that play a theoretical role. And he is clear about what virtues such posits and the theory they are embedded in should have: simplicity, familiarity, scope, fecundity, and success under testing (Quine 1955, p. 247).

Nevertheless, there are two interpretations possible for what it means to organize experience, a narrow interpretation, which I at-

---

ered are always *mental* operations; the arguments do not concern the actual drawing of lines and determining of points on sheets of paper." (!) What were being constructed, in any case, were *not* diagrams, or anything perceptible in the ordinary sense of the word. This is why Plato's doctrines about mathematics struck Greeks (in particular) as plausible.

I should add that Tait (1986, p. 346) *seems* to make the same objection I've just made to views that attempt to bring perception into the justification of mathematical practice. (He directs the point against Gödel 1947, Parsons 1979–80, and Maddy 1980, and writes: "[T]he issue is not whether we perceive mathematical objects, but whether our canons of proof obtain their meaning and validity from such perceptions.") Also see Wagner 1982, p. 269, note 8, where the same point is directed against Hilbert 1925, Gottlieb 1980, and Kitcher 1978.

tribute to Quine (although I have no intention of carrying out a detailed exegesis to prove this), and a richer interpretation. Recall that the identification of the epistemic status of mathematical objects with the epistemic status of empirical objects that I have attributed to Quine admits of only one qualification on his part: The sentential vehicles carrying mathematical terms are more remote from sensory experience than their empirical counterparts.[82]

But in practice we require more of empirical posits than we do of mathematical ones – and it is here that a richer interpretation of what it means to organize experience becomes relevant. We generally require that part of what needs to be organized are *our* epistemic practices within science itself. That is, part of what needs to be organized are explanations of when we've made mistakes and *how* we explain those mistakes. The Quinean observation that likely epistemic change varies inversely with distance from the sensory periphery blends the epistemic roles of physical objects and mathematical ones (they do not differ in *kind* epistemically), but in fact this obscures scientific practice: The antics of empirical objects often explain why we make mistakes, and why we were prone to adopt a wrong theory (for as long as we did, say). The causal roles of empirical objects can be used to explain *in a systematic* way why actual yield deviates from theoretical yield in applications.[83] Furthermore, one seems to have to back up one's empirical posits with instrumental practices – one needs to interact causally with them in some way to guarantee that they are empirically real. Not so with mathematical objects. Presence in space and time thus is not merely a metaphysical oddity enjoyed by some posits and not others; it has consequences – epistemic ones.

These observations show, I think, why the Quinean perspective on theoretical objects located in space and time seems epistemically inadequate, and why reliabilist theories about such objects seem desirable. In fact, such epistemic theories can be taken to place a kind of constraint on realism about empirical objects, namely, the constraint that such objects *should* play a role in explaining epistemic blunders.

I want, therefore, to make a distinction between *thin posits*, which play the sort of organizing role that Quine requires posits for, and *thick posits*, which are required to do the additional epistemic work just described.

Now certainly we can agree that theoretical objects in the empirical sciences are thick posits. But why aren't mathematical objects thin

82 Minor though this qualification is from an epistemic point of view, Quine needs it to explain common intuitions about the necessity and aprioricity of mathematical truths. See Quine 1936, p. 102.

83 I sense something similar to this distinction in Kitcher 1984, pp. 74–6. He notes that different sorts of epistemic explanations are available in the case of 'dogs are black' and in the case of 'every group has a unit'.

posits; why can't Quinean epistemic requirements *as they stand* suffice as a justification for mathematics? That is, why can't we claim that the epistemic story required to explain our knowledge of mathematics turns entirely on the fact that mathematical posits of this sort enable us to organize experience successfully, that such posits offer "simplicity, familiarity, scope, fecundity, and success under testing."[84]

Notice that to treat mathematical objects as thin posits solves *both* Benacerraf's puzzle and the epistemic role puzzle. For the epistemic assumptions of Benacerraf's puzzles are *rejected when it comes to thin posits;* once we explain the utility of mathematics, we have done everything needed to explain why we posit them. Benacerraf's puzzle arises from attempting to force mathematical objects to be thick posits even though mathematical practice clearly marks them out as only thin posits. On the other hand, the epistemic role puzzle is also solved because the epistemic differences between mathematical objects and empirical ones are due to the fact that mathematical objects are thin posits, and thin posits have no epistemic role other than the requirement that they exemplify indirect theoretical virtues such as simplicity, familiarity, and so on.

So, at least as far as mathematical objects are concerned, the Quinean approach satisfies, epistemologically. How does it fare when we turn to reference? Unfortunately, not so well.

We again find Quine persisting in his tendency to assimilate mathematical and empirical objects to each other; for example, Quine's doctrine of *ontological relativity* is meant to apply to ontological commitments, wherever they may be found. Ontological relativity essentially raises the same issues discussed in Section 2. Indeed, Quine motivates the position in his 1969b by invoking permutations (proxy functions) and the Löwenheim–Skolem theorem. That method, with its accompanying tendency to treat our referential powers with respect to mathematical objects and our referential powers with respect to empirical objects as identical, is employed again and again by philosophers subsequent to Quine 1969b.[85]

Apart from applying the doctrine across the board to mathematical and nonmathematical objects alike, Quine claims that there is no fact of the matter about what we refer to – reference is "inscrutable." Consequently, it is hopeless to attempt to marshal further resources

---

84 Well, it will turn out that this won't quite do. But not because thin posits don't have a thick enough epistemic role but rather because they *still* have too thick a role. See Part II.

85 The list is fair-sized. Wallace 1979, Field 1975, Davidson 1986a, 1986b, and especially Putnam 1977, 1978a, 1981, 1983a, 1983b, 1983c, 1984, 1986, 1989. Quine's discussion is complicated by the fact that it evolves over time and involves more than inscrutability of reference. I intend to discuss the details elsewhere, and so will not get into it now.

for fixing what terms refer to, regardless of whether they are mathematical terms or empirical terms.

The primary place to recognize the presence of inscrutability is where engaged in radical translation. However, Quine argues that inscrutability also arises at home, and can be seen when one recognizes that nontrivial automorphisms of one's conceptual scheme compatible with all our linguistic practices are available. But he allows that one can simply acquiesce in one's own referential practices in the latter case: 'rabbit' refers to rabbits, whatever *those* are.

Now my purpose is not to launch a wholesale analysis of Quine's difficult views on reference. But let me make a couple of observations before turning to how these views bear on mathematical objects.

Many philosophers regard Quine's views, when it comes to empirical objects, as inadequate. There simply *are* resources available in this case for fixing reference. I allude to, in particular, the causal resources embodied in our technology and sensory apparatus. My own feeling is that this is largely correct: What is going on with empirical objects, theoretical or otherwise, is different enough from the mathematical case to separate referential problems regarding the two kinds of object.[86]

Thus Quine's view, when restricted to mathematical objects, is not quite as compatible with the intuitions about reference discussed in Sections 3, 4, and 5 as we found his epistemic views to be in respect to the epistemic role puzzle. Furthermore, there are Quinean moves that seem to undercut the cogency of the intuitive differences between empirical terms and mathematical terms. These differences, recall, turn on making a distinction between 'A'-mishaps and A-mishaps, a distinction, that is, between confusions over terms and confusions over things referred to by the terms. But Quine is not going to take such a distinction seriously.

Why not? One reason is his denial that there is any genuine distinction between what we believe about A and what the term 'A' refers to.[87] Recall Quine's parable of the neutrino (1960, p. 16). Imagine that one physicist claims that neutrinos have no mass and the other claims they do. Quine thinks no sense can be made of the worry about whether they are talking about the same particle or different ones.

---

86 I intend to discuss my views on this matter further elsewhere. But it is worth adding that a number of philosophers rebel against the tendency on the part of Quine, Putnam, and others to assimilate referential problems with mathematical objects to referential problems with empirical objects in this way. See Field 1989, p. 277, Lewis 1984, Devitt 1983, and Glymour 1982, among others. Furthermore, it should be clear that one cannot take Benacerraf 1973 seriously and at the same time regard mathematical and empirical objects as on a par in this respect. This, of course, is not an *argument* that Quine is wrong; it is an observation of the general perception of (some of) the philosophical profession. For arguments, see the writers cited.

87 Implicit quasi-quote conventions are in force here.

The distinction between 'A'-mishaps and A-mishaps, however, depends on being able to make a distinction of just this sort: We need to be able to distinguish what the terms 'James' and 'John' refer to from what speakers are trying to refer to in using them – and then further, we need to be able to distinguish the different reasons why what speakers are trying to refer to with such terms differs from what the terms refer to. Thus the distinction between A- and 'A'-mishaps turns on making just the sort of prior distinction Quine denies one can make.

But the Quinean rejection just rehearsed is symptomatic of a much deeper divide between the picture of mathematics the discussion here seems to give and Quine's views. Let me single out four connected views of his for particular criticism. I label the views as follows:

(1) The significance of formalization for reference
(2) The priority of theory over the terms contained therein
(3) The immanence of the referential apparatus to the conceptual scheme it is contained in
(4) The inscrutability of reference

I'll be brief, as, to some extent, I will be repeating the analysis already given.

First (1): For mathematicians, formalization definitely has significance. It can resolve confusions, it can be used to design independence proofs, and so on. The study of formal languages also constitutes an exciting and ever-growing branch of mathematics. But one thing it is not used for is recognizing ontological commitments. By this I don't merely mean that mathematicians fail to regiment their theories in first-order languages in order to recognize what their mathematical theories commit them to. Rather, I mean that ontological commitment is due to the introduction of terminology in natural languages, and clarification of such terminology does not require formalization. Instead what we find, as we've seen, is the attempt to construct certain mathematical objects (e.g., complex numbers) on the basis of others. But there is nothing in this that forces the use of formalized systems.

In Section 1, I said that formal languages are (practically speaking) too awkward for the ordinary mathematician to use in research. But something stronger is true. Formalized systems simply are not needed by mathematicians to do mathematics. They constitute a legitimate branch of mathematics only, not an ontological foundation.[88]

On (2): Surprisingly, given the mathematician's obsession with theorem proving, we have found that the pressure of ontological com-

---

88 In my criticism of Quine's reliance on formalization this way, I detect some kinship with Chihara 1990, pp. 6–15.

mitment is located in the term rather than in the sentence (or proposition) in a significant respect. A mathematical term can successfully refer even if the theory it is embedded in is defective in any of several ways: The theory can be wrong, it can fail to be categorical, it can be so primitive that the suggestion that it can pick out anything by its own referential powers is ridiculous.

On (3): This observation is closely related to the one just made. When Quine describes our quantificational apparatus as parochial – when he describes our theory of truth as immanent – one thing he seems to have in mind is that reference, *such as it is,* is fixed by the theory surrounding our terms, including logic. We have just seen reasons to think this is wrong. But something more sometimes seems to be meant – that we do not understand what (mathematical) terms refer to until we have translated the theory surrounding them in an alien conceptual scheme into our own parochial scheme. This does not quite seem to be right either. We have found that mathematical terms can often travel easily across quite disparate conceptual schemes. Number terminology and how it is used sometimes is more significant in the recognition of what terms refer to than number theory, however developed.[89]

I don't want to claim that reference in the mathematical case *transcends* any particular conceptual scheme, for that probably implies more than I am willing to grant. This much seems true, though: Although there is a bias on the part of the mathematician toward theorem proving, this bias seems to be entirely epistemological, and not ontological, as Quine treats it. Rather, as we've seen, for the mathematician mathematical noun phrases seem to carry a lot of the ontological freight, and when they do, it allows the easy transplantation of mathematical reference from one conceptual scheme to another; in this sense the reference of mathematical terms is not *immanent.*

On (4): Finally, inscrutability of reference is just the wrong sort of phenomenon to get here, and its source seems to be that very ontological priority of the sentence over the term we have placed in doubt. Rather than one's mathematical terms not being fixed in what they refer to, the counter-impression that issues from mathematical practice seems to be that fixing the reference of mathematical terms is, on the contrary, a *trivial* matter.

It might be suggested that when Quine says that there is no fact of the matter about reference, and talks about acquiescing in one's own

---

89 Sometimes, but not always. For example, intuitionistic numbers are distinguishable from classical ones – and it seems to be the theory surrounding the terms that make them so distinguishable. Without a picture of what is actually going on here, it is hard to see exactly why and when terms can be transferred from one mathematical context to another and when they can't be. I give such a picture in Part II.

referential scheme, he is acknowledging the claim I have just made. But there are two reasons to doubt this.

First, the doctrine of ontological inscrutability is shown by permutation arguments and these do not seem to imply that reference is trivial at all. Rather, the conclusion to draw from them seems to be just the conclusion Quine draws: Reference is inscrutable. Considerations that issue in the sort of reinterpretation that Quine's proxy functions glory in cannot simultaneously show that reference is trivial.[90] The other indication is that the Quinean doctrine of ontological inscrutability, because of the sort of arguments used for it, is most naturally opposed by attempting to find resources, that Quine has overlooked, which fix the references of the terms. Thus one considers the fact that Quine's argument is given entirely in first-order terms, and so one considers higher-order logic, or one argues that there are causal resources for fixing what one's mathematical terms refer to. None of this is needed if the doctrine is instead that reference to mathematical objects is trivial.

I should repeat, despite these objections to Quine's views, that it is a Quinean insight that we are adapting to our own purposes: Mathematical objects are (thin) epistemological *posits,* and one does not need to tell a *referential* story (e.g., a causal story) about how we refer to (thin) posits any more than one needs to tell an epistemological story about how we know the things we know about (thin) posits. By describing such posits as thin posits, we have discharged our naturalistic requirements.[91]

Let me sum up. So far, there is much in Quine's views I have suggested must be discarded. Broadly speaking, I want to dismiss his sensationalistic views on ontology while retaining his epistemic insights about positing, at least in the case of mathematics. Furthermore, I have given some reasons for reversing the tendency, starting with Quine, but continuing with nearly every contemporary philosopher of mathematics, to assimilate epistemological and ontological practice in mathematics to such practice in the empirical sciences. I believe I can parlay these hints of another approach into a substantial pic-

---

90 Quine does say reference is trivial, but he does not mean what I mean by this phrase. For example, he writes: "Within the home language, reference is best seen (I now hold) as unproblematic but trivial, on a par with Tarski's truth paradigm" (1986b, p. 460). Thus part of why he thinks reference is trivial has to do with his talk of acquiescing in the mother tongue, and part of it has to has with his views about the status of the Tarskian apparatus with its focus on disquotation.

I do not mean this, for the triviality I have been stressing emerges from the intuitions about reference raised in Section 4, which mark out a contrast between mathematical terms and empirical terms; and both one's acquiescing in the mother tongue and the use of the Tarskian apparatus are things one does with empirical terms, as well.

91 In some sense, the insight predates Quine, however. See Carnap 1956.

ture of mathematics that will avoid most, if not all, of the drawbacks that plague present approaches. In particular, we have been given glimpses of the possibility of a view that will give solutions to the epistemic role puzzle and the new puzzles of reference. If a view of this sort can explain all the epistemic qualities of mathematical practice, including its utility, it will eliminate the need to solve either Benacerraf's problem or the puzzle of referential access.

But before going on to do this in Part II, I want to canvass other extant objections to Quine's views about mathematics that can be found in the literature. My reasons are twofold: First, recall that the hint I intend to take from Quine – to treat mathematical objects as thin posits – is one that is nearly universally neglected, as the focus on Benacerraf 1973 makes clear. It is important to determine whether it faces substantial objections or is merely the victim of a (peculiar) oversight. In determining this, one must not merely look to direct responses to this move, for the move may have consequences that are unacceptable. Second, in discussing objections to Quine's views, we will find other constraints on what an acceptable picture of mathematical practice must look like.

§11 Quine's Approach II

Here are several suggested problem areas for the Quinean approach.[92]

(1) Epistemically, for Quine, mathematical truths are on a par with high-level theoretical laws of science. But, as Parsons 1979–80 points out (and Maddy 1990 and Luce 1989 after him), a statement such as '2 + 2 = 4' just doesn't seem to be in that class. It is too obvious.

(2) Justification of mathematical truth, for Quine, is an empirical matter. It draws its source from its indispensability to science. But mathematical justification, and practice, as Maddy 1990 pp. 45–6 and I (Azzouni) 1990, pp. 93–4 point out, seems to be sui generis. Mathematicians do not look to the empirical sciences *in any way* to justify what they do epistemically.[93]

Putnam 1979 focuses on a variant of this issue. Classical mathematics, set theory, for example, goes beyond application in significant ways; Putnam writes:

Quine seems to be saying that science <u>as a whole</u> is one big explanatory theory, and that the theory is justified <u>as a whole</u> by its ability to explain <u>sensations</u>. [But] . . . the idea that what the mathematician is doing is con-

---

92 I draw these objections, with modifications, primarily from the following sources: Putnam 1979, Field 1980, Parsons 1979–80, Maddy 1989, 1990, and my own 1990.

93 Well, of course I need to exempt computer-generated proofs. But these are special cases, which I discuss later (see Part III, Section 2).

tributing to a scheme for explaining <u>sensations</u> just doesn't seem to fit mathematical practice at all. What does the acceptance or non-acceptance of the Axiom of Choice (or of a known-to-be-consistent but <u>not</u> accepted principle like the axiom "V = L" that Gödel once proposed but later gave up) have to do with explaining sensations? (p. 390).

Now it seems to me that, as Putnam's objection stands, Quine has a rejoinder. First, it can be observed that the Axiom of Choice, in particular, was a bad example to give here. For that axiom is widely used in, for example, functional analysis, which in turn is crucial to mathematical physics as it is currently practiced.[94] We can change the example, of course, to the Continuum Hypothesis, say, or focus on the rejection of "V = L" as Putnam mentions in the quotation just cited, but the damage has been done. One now recalls that Quine has often written about the need for simplicity in one's theories (and that one allows a certain amount of additional "adiopose tissue" in one's theory that does not directly relate to application if there is a gain in simplicity); one also realizes that there is no getting away from the fact that what is unapplied set theory today may become applied set theory tomorrow, and we would be quite foolish if we tried to predict which principles of set theory would succumb this way to application and which ones wouldn't.

To make this sort of objection stick against Quine, therefore, it seems one must look to branches of mathematics that are far less empirically applicable than classical set theory is. And, in fact, the best examples come from alternative nonclassical mathematics: intuitionism, multivalued logics, weird developments in abstract algebra, and so on. On the Quinean view, this sort of mathematics is without justification too. But it seems to be as well justified as classical mathematics is.

Indeed, one presumes to sense Quine's unease regarding how to fit this sort of mathematics into his picture, for he waffles on how to treat it.[95] One tendency he has is to treat the stuff as on a par with uninterpreted calculi. Another is to treat it (or some of it, anyway) as studies in the subject of alien conceptual schemes. But *neither* view seems to characterize mathematical practice correctly. It certainly isn't uninterpreted (or at least it doesn't look that way), even if it cannot be interpreted in Quine's favorite set theory – for example, consider category theory, intuitionism, and so on. And mathematicians (apart from some polemical remarks on the part of early intuitionists and Hilbert) do not see alternative mathematics as particularly alien. Indeed, it seems to be mathematics as usual (again, see Maddy 1989 and my 1990).

94 This point may be verified by glancing, for example, into Reed and Simon 1972.
95 See Quine 1984, p. 788, Quine 1970, chapter 6.

Worse, the suggestion that this sort of unapplied mathematics is like an uninterpreted calculus suffers from a serious disanalogy: The mathematics in question may not be axiomatizable. Intuitionistic mathematics *isn't*, and if we all became intuitionists tomorrow, then classical number theory wouldn't be axiomatizable *either*. By this latter point I mean, for example, that one could still use Gödelian methods to augment the axioms of number theory ad infinitum.

(3) The Quinean doctrine that mathematics is revisable seems equally problematic and to be based on the misdescription of mathematical practice mentioned in (2). Mathematical doctrine never seems to be repudiated in the sense that Quine seems to suggest.[96] Mathematics is often ignored (consider quaternions) but never discarded. Even if we all became intuitionists tomorrow, this would leave the body of classical mathematics entirely *intact;* furthermore, there would be no more problem in doing classical mathematics than there currently is in doing intuitionistic mathematics.

Let us take these objections in turn.

On (1): Notice that an epistemic objection is being made here. Quine has a prima facie obligation to explain away apparent epistemic intuitions about statements; he cannot simply ignore them. So, for example, Quine uses the centrality of mathematical truth in our conceptual scheme to explain our reluctance to falsify such truths. But the impression of obviousness is also an epistemic intuition, and Quine is obliged to explain intuitions like that too.

Of course he is not obliged to concede that *apparent* epistemic intuitions actually are *epistemic* intuitions. So he is well within his rights if he chooses to argue that the impression of obviousness one experiences when contemplating '2 + 2 = 4' is due to habitual and repeated application of the truth in an enormous number of contexts. If the rejoinder is made that this doesn't explain the impression that the truth is obviously *true*, then Quine can counter that what is not in dispute here is his claim that mathematical truths are revisable (at least that better not be in dispute, because intuitions of obviousness are too slender a basis on which to mount an argument for unrevisability). Familiarity is the source of the intuition of obviousness here, but that is no guarantee of truth, and this is why the intuition is not epistemically binding.

The apparent impression that this objection has force against Quine, one concludes, is due to the implicit assumption that the cause of the sensation of obviousness must hail from the same circles as the reasons for taking the statement to be true do. But that is false.

---

96 Perhaps certain proofs are repudiated, for example, certain proofs of Euler's. But even this is disputable; I will suggest otherwise eventually.

Unfortunately, despite the fact that my own rhetoric here almost convinces me that Quine has a rejoinder, I don't think this is true. Habitual and repeated applications of simple mathematical truths do not seem to explain the intuition, which seems not to arise with empirical truths that one also repeatedly applies. The reason is that this means the truths involved have wide *scope* in application. As Parsons points out (see Part III, note 41 in this book), this makes them high-level generalizations, which one should be accordingly *less* sure of. Related to this is the impression that separating the intuition into parts in this way does not seem fair to the intuition, because it is an intuition that the truth fits so many contexts because it is obviously true, not that it is obviously true because it fits so many contexts.

On (2): This also poses trouble for Quine, for it seems clear that the sorts of methods available for justification that arise in unapplied nonclassical mathematics are identical to the methods available for justification that arise in applied classical mathematics. The methods I have in mind, of course, are theorem proving and the various quasi-empirical techniques that mathematicians utilize to recognize the possibility of certain theorems – testing of particular cases, for example. One should mention that out-and-out empirical methods such as computer-generated proofs are also available in the case of unapplied mathematics. Thus the Quinean tendency to separate applied branches of mathematics from unapplied branches just seems off base *epistemically*.

This objection affects the notion of a thin posit given in Section 10, for recall that a thin posit was justified by its capacity to help organize experience. Unfortunately, that sort of capacity is not involved in unapplied branches of mathematics.

For mathematics, it seems clear, posits can operate with even thinner epistemic requirements than those we've placed on thin posits. For perhaps posits in the mathematical context have theoretical virtues even when these do not involve application: Mathematical posits could be valuable for organizing mathematics, and here the epistemic justification will devolve on how the posits enable us to do more and better mathematics. On the other hand, perhaps posits in the mathematical context need no theoretical virtues at all. Perhaps virtue, such as it is in this context, is a matter entirely of the isolated mathematical theory and not of its application in any sense. I will describe such posits as *ultrathin posits*.[97]

---

97 I am not changing the definition of 'thin posit' in light of this objection because I believe there are bona fide examples of such whose justification lies in the organization of experience: I have in mind abstract objects other than mathematical ones, such as properties. We have a tendency to dismiss properties when they don't pan

On (3): Once one gets past the fact that initially established mathematical results invariably need cleaning up, one must recognize that mathematics does seem to accumulate in ways that empirical science does not. This may turn out to be a mere sociological point. Perhaps mathematicians treat their dustbins with more respect than other sorts of scientists do. But I suspect not; and consequently it is hard to avoid the impression that Quine has substituted applied mathematics for mathematics and pushed the latter out of consideration altogether.

But this claim of mathematical accumulation is not one that is entirely easy to understand. I evaded a problem earlier when I said that mathematics does seem to accumulate in ways that empirical science does not, for it is not clear that *mathematical truth* accumulates. The reason for this qualification is that the body of mathematics does not seem to hang together in a complementary fashion. For alternative mathematics can be alternative in the logic it is based on – and then it is not clear how we are supposed to say that all of it is *true*. In particular, in what sense are we to say that intuitionistic mathematics is true?

Let me sum up. It is not clear, from what I've said so far at least, that an alternative story to Quine's is forthcoming that can handle these objections. We have seen good reasons for substituting ultrathin posits for thin posits, but if we are truly to reverse the tendency of late to assimilate mathematical practice to practices in the empirical sciences, we will need to explain the sense, still vague, in which mathematics accumulates. That, in part, is what the rest of this book is about.

out empirically, a tendency lacking in the case of proper mathematical posits. This suggests that 'thin posit' as it stands picks something out that requires more epistemically than posits in mathematics do.

PART II

The Stuff of Mathematics:
Posits and Algorithms

If . . . arbitrarily given axioms do not contradict one another with all their consequences, then they are true and the things defined by the axioms exist.

D. Hilbert to G. Frege

§1 Introduction[1]

I start by giving a brief characterization of what I take mathematics to be. Quite broadly, it is a collection of algorithmic systems, where any such system, in general, may have terms in it that co-refer with terms in other systems. I understand such systems to be fairly arbitrary in character; that is, there are no genuine constraints on what systems can look like. Similarly, I understand the co-referentiality allowable between the terms of different systems to be a matter entirely of *stipulation*.

This is not to say that mathematicians don't prefer the study of certain groups of systems to the study of others, or that they don't prefer certain stipulations regarding the co-referentiality of terms across systems to other stipulations: Of course they do, and I have quite a bit to say about how these preferences arise.

I build up this view of mathematics gradually. In Section 2, I present and modify an initial picture of what an algorithmic system is. In Sections 3 and 4, I discuss what restraints the need to apply mathematics empirically places on any view of mathematics. I also discuss truth, for I am not a formalist, at least in one popular way that doctrine is described: The sentences of systems, although syntactically generated objects, are not to be understood as meaningless. Rather, I take them to be susceptible both to interpretation and to talk of truth, and one of the points of these sections is to make clear how this is possible. In Section 5, I turn to Quine's objections to the notion of truth by convention. The reason this topic is pertinent at this point is that the way I take truth to apply to systems is very similar to pre-Quinean views of truth by convention. My purpose in this section, therefore, is simply to show that my view is unaffected by these objections. In Sections 6 and 7, I explicitly turn to the question of how terms are taken to co-refer across systems, and what advantages there are in doing so. It may surprise the reader that truth has been handled here from the point of view of the isolated system, and that co-referentiality across systems arises so late as a topic of discussion. The reason is that the fact that terms can co-refer with terms in other systems is *not* explained in terms of their "being true" of the same objects, although that fact, such as it is, falls out of the analysis given here. This reversal of the order one might have naively expected is a direct result of mathematical posits being ultrathin. Finally, in Section 8, I return to the intuitions and puzzles raised in Part I and show how the view developed here handles them.

---

1 Some of the material contained in Part II appeared in an earlier form as Azzouni 1990. I have not, however, respected either the identity, contours, or content of the original essay here.

§2 An Initial Picture

Until now, we have been focused on various puzzles that mathematical objects pose for reference and knowledge. As a result, I have had little to say about logic except insofar as that topic bears on mathematical ontology. But now I want to shift direction, turning attention to the actual mechanisms by which new mathematics is generated. Recall how I began this book: "Here is a portrait of mathematical practice: The mathematician proves truths." So far this portrait has been used only to foil certain views about mathematical epistemology, but now it is time to move it to center stage.

Here is the strategy. We'll start with what is essentially a *postulate* view of mathematics: One sets down an arbitrarily given set of axioms, an arbitrary set of inference rules, and then derives theorems from them. But how should one view this sort of practice? An early and appealing positivist position was to regard such a set of axioms, and their consequences, as *true by convention*. Quine, in a famous article,[2] buried the view. I have as little desire to raise the dead as anyone, but we will find that the positivistic slant on this matter, although not right in its entirety, is still worth taking seriously.

Two aspects of Quine's desertion of the positivist perspective on this topic deserve comment, because I think they are both mistakes. First, as we've seen, Quine shifts mathematics, both ontologically and epistemically, in the direction of the empirical sciences. Second, he gives reasons for believing that the notion of conventionality as it is supposed to apply to mathematics and logic either doesn't make sense or is vacuous.

My counter suggestion is to argue that there is a perfectly sensible way that the conventionality of mathematics and logic may be understood. Furthermore, provided we are willing to sacrifice the claim that logic is topic neutral, this notion can even be strengthened to *truth* by convention in a sense somewhat close to the original positivistic idea.[3] We will find, in this case, that not only do we have *truth* by convention but that we have *ontology* by convention as well.

If we are reluctant to sacrifice the topic neutrality of logic (and I will argue that the sacrifice involves far less than it seems to), we can still preserve the suggestion that mathematics and logic are conventional by driving a wedge between the conventionality of something and its truth value. All branches of mathematics and logic, in this case, are conventional in the same sense as before. However, the *truth* or *falsity* of particular branches of these subjects *is* an empirical matter

2 Quine 1936.
3 We eschew, however, any attempt to explain this notion in terms of linguistic meaning. Rather, the operative notion will be that of *algorithm*.

just as Quine claims. In particular, the truth or falsity of a particular branch of mathematics or logic turns rather directly on whether it is applied to the empirical sciences.

We will find that not sacrificing the topic neutrality of logic does not give us as appealing a picture as we would have otherwise, but even in this case I will show that the empirical impact on the truth values assigned to mathematical or logical sentences is neither as significant epistemologically as Quine has thought nor does it corrupt the rather crisp distinction we would otherwise see between mathematics and the empirical sciences.

Let us start, then, given the hint that "the mathematician proves truths," by attempting to explain the practice of mathematics and logic in the most direct way possible. A *postulate basis* has (a) a recursive set of axioms (possibly null) to be used as premises in derivations and (b) a recursive set of inference rules to be used in justifying the steps in derivations. We say that a sentence A *is derivable from* a postulate basis PB ("$\vdash_{PB}$ A"), if there is a derivation of A from the axioms of PB using the inference rules of PB.

Notice that PB also imposes a deductive structure on the entire language it is couched in (or, for the sake of brevity, "PB defines a *deductive relation*"). We say of two sets of sentences S_1 and S_2 that S_2 is PB-derivable from S_1 ("$S_1 \vdash_{PB} S_2$") if, using PB, the sentences of S_2 can be derived from those of S_1. Of course, if we allow S_1 to be null, the deductive relation defined by PB fixes which sentences are derivable from PB. I will continue, however, to mention explicitly the sentences so derivable despite the redundancy, because, although logicians are traditionally concerned as much with the sentences derivable from a postulate basis (e.g., the logical truths) as they are with the deductive relation defined by postulate basis, mathematicians are usually not concerned with the deductive relation but only with the sentences so derivable.

The *system generated from a postulate basis PB* is (a) the set of sentences derivable from PB and (b) the deductive relation defined by PB.

In our discussion of systems to follow, we shall sometimes stress the sentences derivable, speaking of "the sentences that hold in a system," and sometimes we will stress the deductive relation; the choice in stress will depend on what is pertinent to the issues being discussed.

A caveat: The nicest systems are ones in which the deductive relation defined by a postulate basis is coded in sentences that are derivable from that basis. For example, consider those systems that are generated by a postulate basis containing (one or another) standard set of rules for the first-order predicate calculus. In such systems, where Γ is an arbitrary set of sentences, A and B are any sentences, and $\rightarrow$ is the material conditional, we have the result: $\Gamma, A \vdash B$ iff $\Gamma \vdash (A \rightarrow B)$. But, generally, results like this need not hold, and so it is important to include explicitly in the system generated by a postulate basis not only the sentences deducible from that basis but also the deductive relation defined by that basis.

Now it is easy to see that the same system can be generated by many different postulate bases. (Consider, as an example, the various formulations possible of the standard predicate calculus.) Not surprisingly, therefore, it is general practice to abstract away from particular postulate bases, focusing only on the system generated. We shall often follow suit. However, we do not usually have the elements of a system simply at hand; we have to *derive* them in some way. So, for epistemic purposes, it is equally important to be aware of which postulate bases are being used.

With this in mind, call a *postulate-based system (pb-system,* for short) a system paired with a class of postulate bases that generate that system. Although we can only work with pb-systems (for we need one or another postulate basis to enable us to get a grip on a system), we do not normally take ourselves to be committed to a pb-system but rather to the system independently of any particular postulate basis. An indication of this is the fact that when a new postulate basis for a system (already studied) is invented, it is simply added to the kit of postulate bases already available without much fanfare.

On the other hand, observe that when comparing systems, we often notice that (some of) the rules in the relevant postulate bases are held in common. This is to be explicitly aware of pb-systems. But we also do something a little more subtle. Consider the two systems, first-order ZFC and first-order PA. In describing these systems as first order, we are explicitly noting that they are both based on the standard predicate calculus. But, as just mentioned, there are many sets of rules that can be used to generate the logical truths in the language of first-order ZFC and the logical truths in the language of first-order PA; so it is not simply a matter of comparing the postulate bases of two pb-systems. Rather, we are considering subsystems of the two systems and identifying them while disregarding the details of how they are generated.

It should be clear from our discussion that it is common practice to slide from considerations about pb-systems to considerations about systems in abstraction from the posulate bases used to generate them, and vice versa. We will do something similar: I will generally (although not always) speak of systems in what follows, and sometimes it will be clear that I have pb-systems in mind. Context, I trust, will make clear what is going on in particular cases.

Let us now make some general observations about systems. As I mentioned earlier, I do not intend much by way of restriction on systems. In particular, I do not require that they be couched in a first-order language, or that the postulate basis contain the axioms and inference rules of the first-order predicate calculus. This is because nothing in the practice of mathematics seems to press the first-order language requirement, and certainly mathematics, containing as it does intuitionism and other alternatives, is not restricted to classical logic. In fact, I have not even required that systems be couched in a formalized language at all.

By relaxing this requirement, we shall not get into trouble. The crucial mark of a formalized language is that its syntax is explicitly defined,[4] and it is this explicitness that seems to enable us to carry out a rigorous metamathematical analysis of such a language – that is, the definitions of 'theorem', 'proof', and 'truth' for such a language that are the foundation of Gödelian and Tarskian results.

However, it does not follow that without such explicitness, such results do not hold. For example, it is not the case that number theory, when couched in an informal medium, is complete. Rather, the right

---

4 Tarski (1983, p. 166) mentions two essential properties of formalized languages: (a) "a list or description is given in structural terms of all the *signs with which the expressions of the language are formed*," and (b) "among all possible expressions which can be formed with these signs those called *sentences* are distinguished by means of purely structural properties."

I am not relaxing condition (a), for I am willing to consider the vocabulary of natural languages at a specific point in time as fixed. I am relaxing (b). One might wonder whether it is possible for there to be a recursive language for which no structural definition of "sentence" is available. It seems like the answer should be "yes," for all that recursiveness requires is that there be a mechanical procedure for recognizing whether or not a sentence is in the language, and such a mechanical procedure need not give us much information about how such sentences are *structurally* distinguished from other strings couched in the same alphabet. For example, consider some canonical axiomatization of the first-order predicate calculus, and suppose that we consider the set of validities provable from this set in deductions of two million or fewer steps as the sentences of our "language." Such a language is recursive, but the mechanical procedure we have been given for exhibiting it tells us nothing about the structural properties of the sentences. Indeed, it is unlikely that these sentences can be characterized "structurally."

Unfortunately, it is not entirely clear what a "structural" definition is. Tarski illustrates the idea this way: "The general scheme of this definition would be somewhat as follows: *a true sentence is a sentence which possesses such and such structural properties* (i.e., properties concerning the form and order of succession of the individual parts of the expression) *or which can be obtained from such and such structurally described expressions by means of such and such structural transformations*" (1983, p. 163). Because of the second condition here I am not sure any recursive languages are excluded by this definition. I cannot pause to consider possible definitions of "structural definition," and will simply concede that I regard natural languages as formalized if the (implicit) possession of a mechanical procedure for recognizing whether a string in the alphabet of the natural language is sufficient for having a structural definition of that language in this sense. (All italics in the quotations above are Tarski's.)

view to take here is that informal mathematical results can always be couched in one or another formal system without loss to the mathematical content.[5] Indeed, one point of working extensively within *PM*, *ML*, and other systems has been to show that informal mathematical results can be preserved this way. All that seems lost is the appearance, to some philosophers, of categoricity; and we have already argued that the impression that informality preserves categoricity is an illusion.[6]

What *is* crucial to the language that a system in our sense is couched in, therefore, is that its syntax be characterized fully enough that the algorithms applied to generate the sentences of the system can be defined, for this is what marks out the mathematician's grasp of mathematical proof in the system. So, for example, even though arithmetic was not formalized until late in the game, still it would be easy to mark out, *at any particular time,* what numerical algorithms were in play that enabled the mathematicians to generate the results they got. This is why I have required that the postulate basis and the inference rules of the system be recursive, for one does not engage in theorem proving with nonrecursive sets of axioms.

It is clear that from a formal point of view I understand 'system' very broadly. For example, consider that language made up of the mathematical vocabulary utilized by Euler at a particular time, including the notation of infinite series. And consider the rules he allowed himself for manipulating these things. This constitutes a pb-system.

Several caveats: First, this is not how Euler saw what he was doing; and generally no mathematician sees things this way. Mathematicians invariably have a semantic interpretation in mind. As a result they have no compunctions about augmenting systems at any time on the basis of this interpretation – that is, introducing new vocabulary to the language the system is couched in, or new rules to the postulate basis – even if such changes are not conservative with respect to the previous system.

Second, one should not understand a pb-system as invariably containing clean and generally applicable inference rules. That sort of system emerges only at the end of a practice, as the Euler case again makes clear. Instead, one has inference rules with ad hoc restrictions, and in fact sometimes a good case can be made that the system used by a particular mathematician at a particular time is an inconsistent one. In such instances the inconsistency of the system has not been explicitly noted, and when it is, the system is modified, sometimes

5 "can always be . . ." here is highly idealized, however. See Part III, Section 2, for details.
6 See Part I, Section 2.

with a more or less satisfactory ad hoc improvement. (For example, "Don't apply certain rules in these particular cases.")

Now, one might have hoped that all mathematical and logical practice can be carried out in one canonical system, and my mentioning *PM* and *ML* earlier might look like a hint in that direction. But mathematical practice doesn't seem to operate this way. Rather, it looks like in mathematics one is relatively free to choose a postulate basis at will and see what sort of system it generates. By no means need the set of sentences be taken to be "postulates" in any canonical sense, that is, they might just be the latest results one has read in a journal plus some other sentences (sentences specifically of classical logic, say, or other background mathematical assumptions adopted for the purpose at hand) used to generate new results. Actually, one may choose any class of (mathematical) sentences and draw conclusions from them (using some set of inference rules or other).

Of course, if one wants to get noticed in the field, it is important to choose sentences that are currently being taken seriously (regarded as "true" at least, although not always) and draw results from them, or choose sentences from which interesting classes of sentences that are already being taken seriously follow in an interesting way. But perhaps this is a mere sociological point about what sorts of things people find interesting. For the moment, let's assume it is.

This, then, is our initial picture. Mathematics is made up of a collection of systems. Systems are algorithmically generated, and they are *arbitrary* in a rather precise sense: There are absolutely no restraints on them except that their postulate bases be recursive. In particular, syntactic consistency is *not* a requirement, for there are certainly systems that contain every sentence (couched in the language of that system). Such systems are not particularly interesting once they are exposed for what they are, but so far, that is their *only* drawback.

So far, too, there is a precise sense in which systems are conventional. Conventionality arises because there is a total latitude in what rules and postulates may be chosen (apart from the requirement that the set of inference rules be recursive and that there be a recursive set of postulates that the rules can be applied to), and there is total latitude in the vocabulary as well. Calling mathematics and logic conventional, given this picture, therefore, is to draw attention to a property they share with rule-governed *games*.[7] Similarly, there is also a clear sense in which we may say that a system is conventionally *defined,* as the postulate basis is definitive of a system.[8] Finally, systems are, so

---

7 Resnik (1980) discusses Thomae's attempt to treat arithmetic as a "game" played with symbols and Frege's responses to this move. See pp. 56–65.
8 Nothing rules out something being conventionally definable in more than one way.

far, individuated fairly tightly: Change the vocabulary, or change the postulate basis in a way that changes the set of sentences generated or the deductive relation defined, and the system is changed as well.

At this point we can introduce a notion of ontological commitment. Because our systems contain sentences, we can recognize the presence, if any, of noun phrases, proper names, and quantifiers. Let us describe a system Γ as *committed* to an object α if an assertion of the existence of α (an existential assertion) occurs among the sentences of Γ. And let us describe all such objects α as the *systemic commitments* of Γ.[9] Notice that systemic commitments are contextual in a peculiar sense: They are linked to the system they arise from and no sense has been given to the idea that two different systems can be about the same objects. Indeed, commitment here is quite analogous to what goes on in games. In chess, one has a "king," and one has "kings" in card games as well, but there is no sense in which these kings are the same. What we have are *posits* in the narrowest sense of the word – the terms make sense only within the context they are employed in.[10]

A word about the languages that systems are couched in: For the moment, it doesn't much matter whether we individuate languages so that each system is couched in its own language, or we individuate them so that more than one system can be expressed in a language. However, shortly, when we consider combining systems into conceptual schemes, and when we consider applied mathematics, languages in general will contain more than one system. It is important, when languages are taken this way, that we do not set up the language and

9 I admit that this is vague, but because I'm not restricting systems to first-order languages, I have no choice. I should say that *when faced with a formal system* my own instincts on this score tend to be fairly Quinean – I often do not trust the purported systemic commitments of a system until I have translated it into the first-order predicate calculus. But as the discussion in Part I, Section 10, indicates, I want to leave the notion of ontological commitment at a more intuitive level. Certainly mathematicians seem to have no problem recognizing when a mathematical theory commits them to a new sort of object – and the way they do it is simply by recognizing what existential assertions involving noun phrases there are. I think, for our purposes, this is all that is necessary.

I should add that if we restrict ourselves to the first-order case, and eliminate proper names in good Quinean fashion, what emerges is simply the Quinean criterion for ontological commitment. I point this out because we will eventually find this criterion, even in the broad form I've given, inadequate for mathematics.

10 So even when two systems belong to the same language (syntactically speaking), we do not understand the sentences in them as having the same content, that is, if the same noun phrase occurs in two sentences in two different systems, we take that noun phrase to be referring to different things in each case. So, although we identify *sentences* syntactically, we distinguish the *statements* they express. At the moment we have in place a requirement that two tokens of a sentence express the same statement only if they appear in the same system. But eventually this requirement will be relaxed. See Section 5.

Notice that 'statement' as it is used here is a term of art. It is simply an equivalence class of sentence tokens, modulo systems.

THE STUFF OF MATHEMATICS 87

the rules of these systems so that systems interfere with each other. This is fairly easy to do in practice, of course, for where the rules used are the same, one can just allow them to apply across divergent terminology (e.g., as one traditionally does with standard logic), and where such rules are not the same, one can hedge their application terminologically.[11] I will rarely be explicit about this in what follows but, rather, leave it to the reader to make the necessary adjustments.

Let us now return to the issue of posits, and notice that although we have individuated rules so that they may appear in more than one pb-system, we are not allowing posits to so appear. There is no inconsistency in doing this. For example, consider two pb-systems that contain among their rules the axioms of PA. Just because (some of) the rules in these pb-systems are the same doesn't mean we have to assume the objects referred to by the terminology in these pb-systems are the same, and for the time being we won't assume this.

Now, eventually we will find that our denying that ontological commitment is intersystemic won't do for mathematics and logic, for, unfortunately, the analogy between these subjects and games is not quite precise, as considerations in Part I already make clear.[12] But before turning to the drawbacks of this analogy, let me focus on its philosophical advantages. When we modify the picture to get something truer to the exact sciences, our aim will be to preserve these philosophical advantages; so we do well to keep them in mind.

In a certain sense, the establishment of a system is an a priori matter. One merely sets up the rules of the game and then operates with them. Notice also the ontological commitment here is a mere product of grammar; when utilizing a system Γ, one "commits oneself" to the systemic commitments of Γ. But there are no epistemic problems because the adoption of a system and its commitments is arbitrary.

On the other hand, notice that genuine epistemic problems *do* arise with the executing of algorithms. Depending on the algorithm, it can be quite hard to determine whether a given sentence belongs to a system. This is a syntactic restriction on our epistemic powers: If a set of sentences is recursively enumerable but not recursive, then it is

11 Here is a cheap illustration. Consider a language that contains both intuitionistic and classical connectives. Introduce two sets of atomic sentences (I-atomic sentences and C-atomic sentences), make sure the connectives can be distinguished from each other, and set up the syntax of the language so that no sentences containing both intuitionistic and classical connectives are possible, and so that no I-atomic sentences and C-atomic sentences can appear with classical connectives or intuitionistic connectives, respectively. Finally, write the intuitionistic and classical rules so that they apply to intuitionistic and classical sentences, respectively.

For our purposes, how languages are to be individuated is something of a slippery matter.

12 I am alluding principally to the discussion in Part I, Section 4, which indicates that mathematical posits are often identified across systems.

impossible to design a general method that will determine, for an arbitrary sentence couched in the language of that set of sentences, whether or not it belongs to that set. Furthermore there are always computational limitations on our capacity to derive theorems. And this is all as it should be.

We find, therefore, that the primary epistemic obstacle normally faced in the practice of mathematics is present in our picture: The hard work is involved in determining which sentences belong to which systems. In addition, insofar as what is required is the derivation of existential assertions, we find that the *establishment* of the ontological commitments of a system may be very hard – but such commitments simply follow from what sorts of statements are derivable in the system. Thus there are no further epistemic considerations relevant to our knowledge claims about the objects posited by a system and what properties hold of them. In particular, any reliabilist theory that accords a substantial role *to* such objects in our knowledge-gathering methods for learning *about* such objects is simply out of place: To require such a theory is to commit a sort of category error.

So much for the epistemic puzzles. What about the referential ones? I think it is premature to discuss the referential intuitions I raised in Part I; we should wait until the full picture is in place before evaluating it against them. *But* notice how the triviality of the sort of ontological commitment available here obviates fears of ontological relativity. For there really is no sense in which the objects postulated exist apart from their role as referents of particular grammatical structures (e.g., noun phrases) in systems. Indeed, the idea of permuting objects and referential relations here (so that, say, 'A' refers to B and 'B' to A) barely makes sense. In a sense, both A and that 'A' refers to A are stipulations. To consider the possibility that 'A' might not refer to A but to B is to forget how A "came into being" in the first place.

Thus, if this picture sufficed to explain mathematical practice, it would be clear in what sense mathematical posits were ultrathin, why the objects themselves play no role in our story of how we know about such things, and why reference to them is a trivial matter.

§3   Application and Truth

No story about our mathematical practices is complete without an examination of two related aspects of these practices. First, talk of mathematics is interspersed with talk of truth, and second, mathematics and logic are routinely applied in the practices of the empirical sciences and in our daily lives. This requires our philosophical story to explain how mathematical vocabulary can mix freely with the vocabulary of the sciences and everyday life; and it requires the story to tell us what the truth-value status of the group of mongrel sentences that results is. These constraints are well known, and it is also well known that they seem to rule out certain otherwise appealing philosophical moves.

Before turning to a description of these moves and how they are apparently ruled out, I need to get clear exactly what 'application' and 'truth' mean here. That is the burden of this section: I take up 'truth' first and then dedicate a brief paragraph to 'application'.

Talk of 'truth' often seems to many philosophers to involve substantial metaphysical concerns, in particular the "correspondence theory of truth" is often taken in this way.[13] Nothing this strong is needed for our purposes. I want to understand talk of 'truth' as requiring a device whose primary function is to allow us to assent to and dissent from groups of sentences that, for one reason or another, we cannot explicitly mention. Traditionally, this is done by introducing a one-place predicate 'true', which is constrained by Tarski's Criterion T (i.e., the general constraint exemplified by " 'Snow is white' is true if and only if snow is white."). The introduction of this predicate enables the wordy among us to say, " 'John is running' is true," rather than 'John is running', but the value of the device really comes in when one needs to say things like, 'Everything Jack said yesterday was true', or make general remarks about inferential connections between classes of statements, as in, 'If statements of that sort are true, then statements of this sort are true'.

There are formidable problems with the formal construction of such a truth predicate when the language is, as it has come to be called, self-referential, that is, when the truth predicate is supposed to range over sentences containing instances of the truth predicate itself.[14] We will not be concerned with such cases here but only with a sort of case similar to the kinds of cases Tarski concerned himself with. Given a language $\mathscr{L}$, how do we construct a metalanguage $\mathscr{L}^*$ containing (pretty much) the resources of $\mathscr{L}$, but with the addition of a truth predicate that operates over the sentences of $\mathscr{L}$? In particular, given a system in our sense, we would like to know that we can construct a

13 Sometimes the correspondence theory of truth (understood in a philosophical rich sense) is taken to be captured by Tarski's analysis. See Popper 1965, pp. 223-4, Hempel 1965, p. 42, and Carnap 1949, p. 119, for examples. There is some evidence that Tarski, too, saw things this way early on. See Tarski 1983, p. 153. But recently this interpretation has been more or less discarded, and Tarski himself suggested otherwise in his 1944, p. 362. See the following for further discussion, and also Putnam 1978b and Soames 1984.

14 The literature is rich with alternatives to Tarski 1983, the point of these alternatives being that they are formal languages that contain, more or less, their own truth predicates. See Kripke 1975, Herzberger 1982, Barwise and Etchemendy 1987, and my 1991 for examples. Other papers may be found in Martin 1984.

"metasystem" similar in spirit to the system itself. If a system is intuitionistic, we would like a metasystem that is intuitionistic; if the system has certain set-theoretic resources, we would like the metasystem to have similar resources (excepting, of course, whatever additional resources the truth predicate and its fellow travelers require).[15]

Anyone who has read Tarski 1983 may harbor doubts that this is possible. There are two issues. First, recall that Tarski's metalanguages have considerable power. In pointing this out, I am *not* just alluding to the fact that the metalanguage the truth theory for an object language must be couched in needs to be strictly stronger than the object language. I am also pointing out that Tarski restricted his considerations to metalanguages containing classical connectives and classical quantifiers. It is simply unclear, given Tarski's approach, that anything of this sort can be done with fewer resources than the ones Tarski avails himself of.[16]

Putnam[17] seems to suggest otherwise. His claim, in particular, is that what the Tarskian truth predicate captures is fixed by the interpretation of the logic of the metalanguage; and by exhibiting a homomorphism (due to Gödel) from classical sentential logic into intuitionistic sentential logic, he illustrates how an intuitionistic interpretation of classical logic (and presumably, the set theory utilized in the construction as well) is available. But this example does not inspire confidence given *our* concerns, for it does not show that disquotation devices are directly definable in metalanguages with alternative logics. To show that Tarski's device really carries no deep assumptions, one needs to show that it can be made available wherever only slender resources can be found. In point of fact, this seems *false* of Tarski's approach.

Turning to the object languages, recall that Tarski does not give methods for constructing metalanguages that are applied to every

15   The reader should not be alarmed by the switch from talk of 'metalanguages' to talk of 'metasystems', for the switch is purely rhetorical. I define neither term that precisely. What is at issue is how one is to blend talk of truth with talk of systems. This shifting in terminology is helped by the fact that my use of 'language' is somewhat slippery, as I've already noted.

16   This is relevant to the question of how philosophically neutral Tarski's approach really is. Arguably, intuitionists or nominalists should worry about using it. Notice that applications of Tarski's methods to alternative logics invariably take place using classical resources. For example, possible-worlds semantics *is* the translation of modal idioms into a classical metalanguage (where talk of necessity and possibility is eliminated in terms of classical talk in the context of an enriched ontology). Similarly, the model theory for intuitionistic logic is carried out in a classical metalanguage. Generally, the Tarskian project is seen as one of giving a "model theory" for a logical system, that is, a characterization of a class of (set-theoretic) objects (analogous to standard models) that can be used to assign semantic values to sentences of the language the nonstandard logic is couched in. Talk of truth *simpliciter*, in turn, become essentially a specification of a particular model or class of models (e.g., the *actual* world).

17   Putnam 1978b, pp. 25–30.

sort of object language possible. The object languages he considers are of particular forms – for example, they invariably contain standard connectives. No indication is given how to generalize the approach to alternatives that differ strikingly from the cases Tarski considers, and indeed the subsequent history of semantics seems to show that it simply is not obvious how one should extend Tarskian methods to other cases.[18] Part of the problem is that when one utilizes classical logic in the metalanguage as Tarski does, it is nontrivial to provide an interpretation in these terms for nonclassical connectives. A simple homophonic translation is not available.

Systems, recall, can vary wildly in their resources. Even granting that Tarski's approach is philosophically neutral in that it does not presuppose a particular philosophical stance toward substantial notions of truth, still, it does not seem to be technically neutral. If we consider a system based on a nonstandard logic, and with meager mathematical resources, it seems that any introduction of a disquotation device will call for the importation of additional (and powerful) logical idioms not in the spirit of the object language under study; and their application to the language will be a nontrivial matter.

Following Leeds (1978), recent philosophers have recognized that grafting a truth predicate onto a language calls for, at most, a device that can express infinite conjunctions and disjunctions.[19] Such a device, it is clear, is *easy* to add to a language with arbitrary logical properties, and such a technical fact has helped to convince some philosophers that truth really is philosophically neutral – something that Tarski's theory on its own did not quite succeed in doing.[20] Field, in fact, notes that a truth predicate can be defined in terms of substitutional quantification (where sentences are the substituents), and argues that because the substitutional quantifier is best understood as a device for expressing infinitary conjunctions and disjunctions, and the latter devices are no more to be interpreted in terms of sentences than finite conjunctions and disjunctions are, adding a truth predicate this way does not introduce new ontological commitments (to expressions, say).[21]

I need to make several points at this juncture. First, a language with the general capacity to express arbitrary conjunctions and dis-

---

18 Consider again the *nontrivial* extension of Tarskian methods to intuitionism and modal logic.
19 Leeds writes: "Truth is thus a notion that we might reasonably want to have on hand, for expressing semantic ascent and descent, infinite conjunction and disjunction" (1978, p. 121). Putnam writes, "If we had a meta-language with *infinite conjunctions* and *infinite disjunctions* (countable infinite) we wouldn't need 'true'!" (1978b, p. 15). And more recently, Field writes: "*A disquotational truth predicate serves as a device of infinite conjuction*" (1989, p. 244).
20 Well, when it did, it did because philosophers thought nothing more was presupposed in it than is presupposed in simpler devices. See Putnam 1978b.
21 One also finds this claim in Grover 1990, p. 224.

junctions is invariably stronger than a language with substitutional quantifiers or one with a truth predicate.[22] This leads to our next point: Although all three of these sorts of devices are going to require infinitary inference rules, the *sorts* of rules required need not be the same. This is important because if we are going to admit infinitary inference rules, we want ones that admit of a tractable (finitary) notation for the resulting proof theory, and this simply is not forthcoming if arbitrary conjunctions and disjunctions are admissible.

Let me stress that our desire for a tractable proof theory is not necessarily a concern of the other philosophers that have been mentioned here.[23] We are concerned with this because the project on the table is to find some way of combining talk of truth with the picture we have recently presented of systems. That is, we want metasystems, systems that give us inference rules and axioms (if necessary) for the truth predicate present.

This motivates the approach I give in the appendix.[24] Let me take a few minutes to sketch it out, although I leave the drudgery of the details for the appendix itself. Consider any system couched in a language $\mathcal{L}$. Our approach will be to augment this system with a truth predicate, a quotation device, and substitutional quantifiers.[25] We

22 Leeds (1978, p. 128) notes that a finitary notation *is* available for infinite conjunctions and disjunctions if, for example, they are recursive; this suggests that he is aware of the fact that much less than the capacity to express arbitrary infinite conjunctions and disjunctions is needed to serve the role of a truth predicate. Similarly, Field notes that substitutional quantification captures only "highly regular" infinite conjunctions and disjunctions, and thus that the truth predicate is definable with only the capacity to express such infinite conjunctions and disjunctions available in a language.

23 In particular, Field is concerned more with what ontological freight disquotational devices carry than he is with what the proof theory looks like, although he does note that substitutional quantification is a finitary notation. Similarly, Leeds makes the observation, already mentioned in note 22, about the notational tractability of recursive infinite conjunctions and disjunctions. But neither of them pays any attention to the question of what the resulting proof theory will look like. For example, if we are required to admit derivations of arbitrary countable ordinality, then we will find ourselves face-to-face with a system that is rather user-unfriendly.

Interestingly, neither Leeds nor Field mentions infinitary inference rules. The reason may be that they are both aware that strong completeness can be regained in the context of substitutional quantification without introducing such rules either by allowing term extensions or term rewrites in the sense of Leblanc 1983. Unfortunately, neither move makes much sense in a case where the substitution instances of the quantifier are *sentences* rather than names.

24 Motivates the approach, but doesn't give reasons for *requiring* it. Any approach that provides a disquotational device is acceptable, provided only that the result is a system. I certainly don't want to restrict systems by requiring that when a truth predicate is present, it must be one of the sort I sketch in the appendix. Rather, I am only out to show what is *possible* here.

25 I have been speaking all along of introducing the truth predicate for a language in a metalanguage (or metasystem). But actually, a metalanguage is unnecessary, provided we restrict in one way or another the scope of the truth predicate. That is the approach taken in the appendix.

then augment the derivation rules available in the system with additional rules for these quantifiers, the truth predicate, and some other minor apparatus. We do not add new connectives, but adopt the connectives already available in the system (if any). We shall find, of course, that an infinitary inference rule is needed. But, as the results in the appendix make clear, the infinitary aspects of the system are so tame that any infinitary derivation can be mimicked by a finite one (using some simple defined notation).

It is easy to show that these devices supply us with everything we need from a truth predicate. But they do more. Under certain circumstances (depending on how rich our capacity to describe the language is) they are sufficient for supplying us enough power to describe the semantics of the language too.

Due to the generality of the situation we are working in, these devices are not always conservative. One might have hoped, that is, that introducing these devices into any system Γ would result in a conservative extension Γ^* of Γ; that is, if $\vdash_{\Gamma^*} S$, and S contains only vocabulary from the language of Γ, then $\vdash_{\Gamma} S$. In particular, we might have hoped that if the original system is syntactically consistent, then so is the new system. But in general this is false. For example, it is possible for there to be systems with inference rules that apply to sentences of the syntactically simple form, predicate followed by constant, but don't apply to more complicated sentences. Introducing this apparatus into such a system is a way of applying such inference rules to the more complicated sentences via the assertion of their truth. If the system is bizarre enough, adding the apparatus of the appendix can make a syntactically consistent system of this sort syntactically inconsistent. Of course ad hoc maneuvers are available for preventing the introduction of a truth predicate from making a syntactically consistent system inconsistent. I forgo details.

This suggests the following moral: In cases where this sort of truth predicate is not conservative, it is not that the concept of 'truth' in some sense has unexpected derivational clout; that in some sense by adding a truth predicate, we are introducing a notion with powerful implications. Rather, where the result of adding it is not conservative, the situation it has been introduced into is derivationally pathological.[26]

I should also add that introducing both the substitutional quantifier and the truth predicate may seem redundant because, given the substitutional quantifer, the truth predicate is definable. But this is true only if certain logical resources are available, such as something like the material conditional. In the general context I am working in, I cannot assume I have anything like this.

26 I confess that talk of 'pathology' here is just rhetorical fluff. For from the point of view of the system, the introduction of the truth predicate is not conservative; the truth predicate, even our sanitized version, is not always neutral. Nevertheless, it

Let me sum up. We take talk of truth to be talk best captured by a (more or less) neutral disquotation device applicable to a wide range of systems. The device is neutral at least in the sense that its adoption does not commit us to particular logical devices (other than substitutional quantification and a certain set of derivation rules). It is not neutral in another sense, of course, because its adoption can prove *not* to be a conservative addition to a system under certain circumstances. I will eventually show, however, that it is philosophically neutral nevertheless: The use of the idiom, despite its failure to be conservative, does not commit us to any particular ontological position with regard to mathematical objects.[27]

Let us conclude this section with a quick discussion of the concept of 'application'. Here, too, I have something philosophically thin in mind. Let us assume we have a nonmathematical vocabulary and truths couched in that vocabulary. What is required are two things: first, that we be allowed to use the sentences from the languages of (certain) systems as inference steps from truths couched in this nonmathematical vocabulary to other truths in the nonmathematical vocabulary. And second, that we be able to utilize the vocabulary from (certain) systems in our formulation of nonmathematical truths. That is, that we be allowed to formulate truths mixing the vocabulary of the systems in question with nonmathematical vocabulary, and draw inferences from such sentences using theorems from (certain) systems as well as from other sets of nonmathematical truths we may know.[28]

Notice one thing, therefore, that application of mathematics requires: The ontological content that mathematical sentences have must be carried beyond the system in order to apply such sentences. We put this issue on hold until Section 6.

§4   Systems, Application, and Truth

We turn now to see how the constraints described in the last section affect what philosophical options are available here. In Section 2, we presented a picture of mathematics in terms of systems. How are we supposed to blend this picture with talk of truth in a way that respects our need to apply (some of) our mathematics? Here is a simple move.

helps to notice what the factors are that are causing this nonneutrality when evaluating the philosophical significance of the truth predicate – in particular when evaluating what connections it has, if any, to particular philosophical positions such as mathematical realism.

27 See Section 5.

28 Field 1980 has attempted to show that we can relax the second requirement in certain restricted cases. How successful this attempt has been is controversial. See Field 1980, 1989, Malament 1982, Resnik, 1985, and Shapiro 1983. At present it seems clear we are still stuck with the requirement.

THE STUFF OF MATHEMATICS

Why not claim that all systems are *true*, that is, that every sentence in any system is true? The problem with doing this straight-out is that *if* the sentences in such systems are all true (and the inference rules are all truth preserving), one can pool them and draw the conclusion that *every* sentence is true. The upshot is that what started out looking like a collection of distinct systems ends up being one single inconsistent system. This result is problematical not just because of the sheer boredom it engenders (although that is important), but also because we intend to *apply* mathematics.

Perhaps we can solve the problem by making sure each system is supplied with its own special terminology and constraints so that the particular inference rules of a system are only to apply to sentences within that system. But, again, such a solution is nixed by our desire to *apply* mathematics. We must be given some way of mixing the terminology of our systems with the terminology from ordinary life and the sciences, and this is blocked if we restrict the operation of theorem deriving only to within systems.

Perhaps we should drop talk of truth *simpliciter* for talk of *relative* truth; perhaps, that is, we should take mathematical statements to be true relative to the systems they belong to. This is the sort of view that assimilates mathematical truth to fictional truth (at least on some stories of how fiction works). But again, we face problems with application. We may derive nonmathematical statements from a mixture of statements from several different systems as well as from statements that aren't mathematical at all. What is supposed to be the truth-value status of such statements?

Two last views, which, prima facie, are awkward because of the above constraints regarding truth and application, are ones that either (a) take mathematical statements to have no truth values at all or (b) take mathematical statements to be disguised hypothetical statements containing in their antecedents the postulate basis of the system (one or another version of what has come to be called *If-Thenism*). Again, the problem here is deciding the truth-value status of statements with nonmathematical vocabulary that are derived from other such statements using pure mathematical or logical truths. Such statements, recall, are traditionally taken *both* to be true and to be categorically asserted.

I should point out that the discussion thus far has not been intended to give a definitive refutation of the earlier suggestions – I am well aware that subtle moves are available to the philosopher of mathematics who is wedded to the positions sketched; indeed eventually I will argue for a version of the first suggestion raised. My purpose so far has been primarily expository: to show how considerations about truth and the application of mathematics and logic outside the domain of the exact sciences bear on what sort of picture we are allowed to give of the exact sciences themselves.

I want now, however, to sketch briefly a classical solution to the problem that is due largely to Quine and to raise problems for it that will motivate my own positive suggestions. But before doing this, I must clarify a matter of terminology: We will be making heavy use of the notion of 'conceptual scheme'. This phrase is one that comes easily to mind whenever discussing Quine's views, and usually one does not worry very much about explaining what is meant. Indeed, I have used the term 'conceptual scheme' before, namely, in Part I, Sections 10 and 11, where I discuss Quine's views; and I felt comfortable enough using his catchphrase that I did not bother explicitly to note it. And again, now that a view due largely to him is on the table, I find myself speaking freely of conceptual schemes.

The sole reason for my current terminological qualms is that I want to continue to use the term even after we reject Quine's views and substitute views of my own where the phrase will not quite play the same role. So I must give a definition and then several caveats to illustrate how I mean this definition to be taken.

I will mean by 'conceptual scheme' simply this: *a collection of pairs, each of which contains a statement in a given language £ and a truth value assigned to that statement.*

Three caveats:

(1) I am giving a definition that is entirely general. In cases where certain logical resources are available, much less can do the job. For example, in the classical case where a negation connective is available that allows us to express our taking a sentence to be false as the taking of its negation to be true, the conceptual scheme can be taken to be a collection of statements (we take to be true).

Even though, in general, what is contained in conceptual schemes are ordered pairs, I will often, for ease of exposition, speak loosely of a set of sentences being contained in such a conceptual scheme. What is meant, of course, is that a collection of ordered pairs, each one containing one of those sentences and its truth value, is contained in the conceptual scheme.

(2) I do not mean such conceptual schemes to be, in general, closed under implication (however that is taken in the particular scheme). Neither, however, are conceptual schemes to be regarded as finite because a great deal of tacit knowledge is often contained in such things; a large number of "obvious" truths can be known by the holders of such a scheme even though they are not written down anywhere (or even thought of explicitly by anyone). Taking the conceptual scheme to be closed under derivation solves *this* problem but obscures epistemological issues I take seriously. What is needed, clearly, to make this

notion entirely rigorous is a substantial analysis of "pragmatics." Unfortunately, what I have said here will have to do meanwhile.

(3) I do not in general mean to identify a system with a conceptual scheme. A conceptual scheme is a much broader sort of object that can contain the parts of many systems. I say 'parts' because, although we are understanding systems as closed under implication (as it is understood in the system), we are *not* understanding conceptual schemes this way. Thus, in the case where a conceptual scheme relies on a standard two-valued logic, it does not follow that every sentence in the language of the conceptual scheme appears in the conceptual scheme.

I should say that I am understanding conceptual schemes in this way because, in some sense, they encapsulate the set of beliefs we (collectively) have at a particular time, and that set can change drastically when we crank out derivations of new statements from the ones we currently believe. To take conceptual schemes as closed under the derivation relation would be to distort seriously the role that logic, for example, plays in our conceptual scheme by presenting all such logical truths as, in some sense, immediately accessible to us. The situation is rather different, of course; over time, we have successively larger and larger subclasses of the truths of logic available to us.

Let me now present the classical solution that I spoke of. A class of sentence schemata is recursively generated. Instances of such schemata, using *any* terminology available to the "conceptual scheme," are taken to be *logical truths*. That *any* terminology in the conceptual scheme may be substituted into the schemata designated marks logic out as *topic neutral*. Thus logical truths take on a special status: They can be applied anywhere.[29]

Having done this, we can classify some systems as false and others as true. For example, any system containing any instance of a logical falsehood is false (we'll understand a system as false if it has any false sentences in it).[30] Other systems still have a shot at being (entirely) true.

But not *all* such systems can be true, because they still disagree on their mathematical statements, and so pooling them will still generate an inconsistent system. For example, there are alternative systems that agree on their logic but disagree on their set theories. In one, we find the Axiom of Choice, and in another, we find the negation of the

29 Admissible inferences from sets of sentences to sets of sentences are marked out in a similar way. I omit details and do not focus in what follows on the deductive relation defined here.

30 Notice that in doing this we are taking sentences in different systems to express the same statements. I eventually discuss what exactly is involved in doing this. Meanwhile, because I am describing an alternative I will not adopt, we simply take this practice for granted.

Axiom of Choice. How do we settle the truth values of these systems? Well, we look to applications. We choose any class of systems that supplies all the mathematics we need for our applications and that does not generate (collectively) any instance of a logical falsehood. In making this choice, matters of mathematical taste, such as mathematical simplicity, intuitive appeal, and so on, may play a role. For example, we may choose ZFC, despite the fact that its resources go quite a bit beyond the needs of empirical science, simply because a neat, smaller system that is intuitively appealing and equally adequate to our empirical needs is not available.[31]

One caveat: In choosing such a class of systems, we have no way of verifying (absolutely) that we have what we want in hand, not at least if it is powerful enough to do the job we need. Thus our choice is always subject to the twin qualifications that we may find that we need more to do the job needed, or that what we have may be inconsistent.

Now by no means has the truth-value status of statements in every system been decided by this approach. Two broad classes of systems have been left out. One is those systems whose logical basis is different from the one we chose because of the applications we needed. For example, presuming our choice of logic is the first-order predicate calculus, left out of consideration so far are the various branches of mathematics built on intuitionistic logic, one or another modal logic, or one or another multivalued logic, not to mention an indefinite number of other possibilities.[32]

The other class of systems that has been left out is that based on the same logic as the one chosen (because of the applications needed), but containing a richer mathematical vocabulary than that available to it. For example, ZFC can always be augmented;[33] and certain augmentations are currently under intensive study by mathematicians. If these are not taken to have supplanted ZFC in our conceptual scheme, what is their status? For the rest of this discussion, I describe the system of mathematics and logic adopted because it best suits the empirical sciences as 'standard mathematics' and 'standard logic', and I call the alternative systems 'nonstandard mathematics', and the logics they are based on 'nonstandard logics'.[34] For illustra-

---

31 Neatness, in this respect, is a matter of taste, and tastes can change. Recently, for example, Quine has been tempted to go with less. See Quine 1990, pp. 94–5.

32 To be accurate, some of these *have been decided* if we take sentences to have the same content across systems: Many are *false*. But this is problematic since such an evaluation is not fair to the spirit of these alternative systems. See the following.

33 I have something fairly simple in mind. We can always add new terminology and new postulates, where the new terminology is not definable in terms of what is already available, and the postulates are not provable from ZFC.

34 I should warn the reader that there is no connection between this use of 'nonstandard' and the common use of the term as it is applied to Robinson's nonstandard analysis.

tive purposes (although nothing turns on this choice) we can take "standard logic" to be the first-order predicate calculus, and "standard mathematics" to be any mathematics that can be couched in ZFC. So our question is: What is the status of nonstandard mathematics?

The approach at hand does not leave us with any satisfactory options, as I shall show by running through them. First, we could take all such systems to be false. The drawback is that talk of falsity and truth has been linked to the class of systems we have adopted. Recall that such talk is involved in a device built on an underlying language; it cannot be grafted onto another language unless the sentences there are already included in our language.[35] Thus, in order to describe mathematical statements as "false," we have to include them within our conceptual scheme.

Now, sometimes, without actually admitting such systems into our conceptual scheme, we license talk of the truth or falsity of statements belonging to those systems by taking the connectives (and other logical idioms) in them to have the same content as certain connectives (and logical idioms) of our own. One says, for example, that '$p \lor \neg p$' is rejected in the intuitionistic system. But this is a problematical claim because it is not clear that the intuitionistic '$\neg$' means the same thing as the classical '$\neg$' or that the intuitionistic '$\lor$' means the same as the classical '$\lor$'. Indeed, generally, it is quite hard to compare sentences in languages with alternative logics on a connective-by-connective basis in a way that takes the meaning of such connectives to hold across the languages. Certainly, when one considers logics farther afield from classical logic than intuitionism is, such a strategy looks less appealing.

Another option is to take all nonstandard mathematical systems to be uninterpreted.[36] But we have *already* noted that this approach has a drawback: From the mathematical point of view there is no difference between applied and unapplied mathematics. In particular, there seems to be *no* problem with the interpretation of unapplied mathematics. So, if we adopt this strategy, we are simply ignoring standard mathematical practice. I should add that another objection to this approach is that it takes a metamathematical attitude toward the alien mathematics in question. Mathematicians who draw results in such branches of mathematics do not seem to have adopted such a stance toward their mathematics.

---

35 This is one thing that is meant by Quine when he claims that the notion of truth is immanent to our conceptual scheme. See Quine 1960, pp. 23–5, and Quine 1986a, p. 316.
36 Quine is often tempted to take this view. See, for example, Quine 1984, p. 788.

There is yet another group of unsatisfactory options. This is to take such alternative mathematics as practices in an alien conceptual scheme.[37] We have three possibilities.

(i) We may claim that we don't understand such talk (say, talk of the intuitionistic existential quantifier) until we have translated it into our conceptual scheme. This is not very appealing because such translations are rarely in the spirit of the original project.

(ii) We may adopt a metamathematical perspective toward such nonstandard mathematics and study them from our own vantage point. The study of intuitionism by Kleene and Vesley 1965 falls into this category, for they allow themselves all the resources of classical mathematics. Indeed, a large amount of nonstandard mathematics is studied in this way, with very interesting results. This option is not quite so bad as the option earlier considered of treating alien mathematics as uninterpreted, for one actually supplies an interpretation for the nonstandard mathematics, although it is one in classical terms. Again, however, there is the objection that much of the mathematics cannot be interpreted this way, for it does not seem to involve the mathematician taking a metamathematical stance. But even if we waive this objection, there is another. This is that not all metamathematical studies of nonstandard mathematics fall into this category; for often, we do not want to allow ourselves any further resources than the mathematics under study allows us. For example, we can study intuitionistic mathematics from a purely (meta)-intuitionistic viewpoint. In that case, our metamathematical studies of the alien subject matter will not be classical either.

(iii) We may treat the practice of nonstandard mathematics as excursions into alien conceptual territory. On this view mathematicians go native, as it were, when they study intuitionism. This, too, simply doesn't seem to be true to mathematical practice. Mathematics doesn't seem to resemble anthropology.[38]

We have canvassed all the possibilities available to us, and I have claimed that none of them is satisfactory. As it turns out, what is blocking a satisfactory solution here are the implicit applications of two philosophical principles: one, a particular version of Occam's razor, the principle that we should eliminate as many ontological commitments as is commensurate with successful science;[39] and the other,

---

37 Quine has also been tempted by this view. See chapter 6 of Quine 1970.
38 This last suggestion is the approach I tentatively settled for in Azzouni 1990, pp. 85–6. But I find what is forthcoming much more appealing.
39 The version of the principle I am thinking of may be found in Quine 1976, p. 264, where Quine writes: "Of itself multiplication of entities should be seen as undesirable, comformably with Occam's razor, and should be required to pay its way. Pad the universe with classes or other supplements if that will get you a simpler, smoother overall theory; otherwise don't. Simplicity is the thing, and ontological

the requirement that logic be topic neutral. In offering my alternative solution, therefore, I will present it in two stages. The first stage is where we relax Occam's razor, and the second is where we relax topic neutrality.

First, what *is* the problem that the application of mathematics poses for us? Simply this: We want to know that the mathematics we apply to the empirical sciences is internally consistent and consistent with the empirical sciences. But this is easily achieved without excluding nonstandard mathematics based on standard logic from our conceptual scheme altogether. All we need do is include such systems in our conceptual scheme, provided that we distinguish them terminologically from standard mathematics, and provided we restrict the terminologies of such systems so that they are not allowed to mix either with empirical scientific (or everyday) vocabulary or with the vocabulary of other systems. Consider, for example, one or another version of nonfoundational set theory based on the standard predicate calculus.[40] We can distinguish it notationally from standard set theory (and other branches of mathematics we study), and we can restrict the system so that it cannot be applied. Because our truth predicate is disquotationally linked to our language, in this case we can regard nonfoundational set theory as true: It is not about the same subject matter as ZFC is. ZFC is about sets, and nonfoundational set theory is about sets* (say). And although set theory is applied, nonfoundational set theory isn't.[41]

It may look like we are substituting for Occam's razor a profligate invitation to multiply entities frivolously.[42] But in doing this I am really only licensing what is already standard mathematical practice. Mathematicians do not forbear inventing new mathematical objects just because doing so might be an unnecessary multiplication of entities. On the contrary (at least in the twentieth century). Nor should

economy is one aspect of it, to be averaged in with others." For our purposes what is crucial here is the methodological move of applying Occam's razor *both* to ontological commitments as they arise in the empirical sciences *and* as they arise in mathematics. My denial of this principle is meant to operate *only* as far as mathematical commitments are concerned.

40 See, for example, Aczel 1987.

41 Notice that once the need to segregate nonstandard mathematics from standard mathematics insofar as their different claims about the (otherwise) same entities is taken care of by terminological modifications – so that now the subject matters of the two disciplines differ – no motivation remains for taking nonstandard mathematics to be false. Indeed, it clearly fits mathematical practice better to treat branches of nonstandard mathematics as true but about a distinct subject matter rather than as a false alternative to the same subject matter. Notice that talk of 'truth' here is understood disquotationally in any case. We assert the sentences of the system in question, that's all it comes to.

42 Actually this is nothing compared to what's coming once we take up the invitation to relax the topic neutrality of logic. See the following.

one assume that it is a standard requirement that new mathematical entities must invest in their own existence either by somehow enabling us to augment mathematical results in already preexisting mathematical domains or by enabling us to provide new and exciting empirical applications. That these are values no one can doubt, but by no means are they the sole or even primary motives behind mathematical inventiveness.[43]

Now the primary motive for Occam's razor is the desire for theoretical simplicity. It certainly is true that in some sense additional ontological commitments can complicate a theory, and they can also complicate collections of theories if we routinely bring these theories to bear on each other and on each other's data. So I certainly can't deny the value of something like Occam's razor in mathematics entirely. For example, there is certainly a gain in simplicity when one is able to take two types of mathematical object and reduce one to the other, so that the theorems that hold of the latter sort of object can be used to deduce the theorems that hold of the former sort of object. And certainly any theory, mathematical or otherwise, may have so many ontological commitments that it becomes, in some sense, intractable.[44] But such requirements do not hold of mathematics generally – that is, one is not required in any way to restrict the ontological commitments of mathematics *as a whole,* and indeed, it is hard to see what the point of such a requirement would be.

On the other hand, something like *this* sort of application of Occam's razor does seem to be at work in the empirical sciences as a whole. One does want to keep the ontological commitments of the empirical sciences *as a whole* to a minimum. One reason for this asymmetry, I think, is that empirical theories do not have *built-in* a priori restrictions on the sort of data that can be brought to bear on them; data for a theory, that is, need not be raw data: They can be what results from the application of *other* empirical theories to (raw) data. Therefore there is no way to rule out how different empirical theories may affect one another. The sciences, in principle anyway, are inte-

---

43 Whole branches of mathematics such as (topics in) abstract algebra and generalized recursion theory exemplify this point. For an almost lyrical presentation of the same claim, see Hardy 1967. For (rather melodramatic) complaints about the negative effect of this tendency on twentieth-century mathematics, see Kline 1980, especially chapter 13.

44 Although it must be pointed out that if a mathematical theory becomes intractable, it may fall into oblivion; but that is not the same as the case in the empirical sciences where a theory may be *rejected.* We have already pointed out (Part I, Section 11) that there is a sense in which no mathematics is ever discarded. This makes it hard to understand how one sort of use of Occam's razor can be applied to mathematics at all.

grated;[45] but in mathematics there is segregation. One may use classical methods to study intuitionism, but there is no sense in which results in intuitionism or in classical mathematics impact on each other. Thus there is good reason to keep an eye on the ontological commitments of the empirical sciences as a whole, and, by the way, also on the ontological commitments of *all* the mathematics that is applicable to the empirical sciences. But no such factor can be at work with pure mathematics generally.

We have used a terminological sleight of hand to bring the practice of nonstandard mathematics, when based on standard logic, under the rubric of our conceptual scheme and into the same language that the rest of our mathematics is housed in. The same sort of move can be used to bring in the nonstandard mathematics that is *not* based on standard logic. For we can also restrict the terminology of such systems so that it cannot be mixed with the terminology of the empirical sciences, that of our daily lives, or with the terminology of other systems. At this point, however, the cost is the topic neutrality of standard logic, for the result of doing this will be that there is terminology in our conceptual scheme that cannot be substituted into standard logical schemata. That is, we can take standard logic to be applicable to the vocabulary of the empirical sciences, the vocabulary of our daily lives, the systems of mathematics that are also applicable to the empirical sciences and to our daily lives (what I have called 'standard mathematics'), as well as to the nonstandard mathematics that is based on standard logic. But that is as far as it extends in our conceptual scheme. It does not apply to the terminology of those systems that are not based on standard logic, and this is why it ceases to be topic neutral.

Notice what this move gets us. We are allowed to apply talk of truth because it is a disquotational device, across all the statements of our conceptual scheme, and the reason the systems do not collapse into one large syntactically inconsistent system is that the disquotational device respects the terminological restraints that are built into the systems and the logics that such systems are based on.[46] Furthermore, we continue to respect the constraints that application places on us. Not any system can be applied empirically, and thus the safeguards (however effective) that we had in place for protecting the mathematics applied from inconsistency are still in place. Furthermore, the pri-

---

45 *In principle*, because in practice it can be quite difficult to bring the results of one science to bear on another. In part, this explains why we have special sciences. I cannot go further into this now, as it would lead to a lengthy digression.

46 Tarski-style truth, based as it is on classical logic, is not prima facie suited to this move. I hazard to guess that this is part of the reason why Quine has never considered it.

mary value of topic neutrality, the easy and simple application of standard logic to the empirical sciences and to those branches of mathematics that are applied, is still intact. All we are doing is preventing the application of standard logic to alternative mathematics, where it is out of place in any case, and arguing that restricting the topic neutrality of logic in our conceptual scheme is better than preserving that neutrality at the expense of excluding alternative mathematics from our conceptual scheme altogether.

What is the value of the topic neutrality of logic? The simplicity that results. But this simplicity must be in respect to *all* our intellectual practices (at least in the sciences), and if it must be bought by excluding from our conceptual scheme much of the practice of pure mathematics, something has gone seriously wrong with our simplicity "calculus." Such mathematical work goes on in mathematical circles in any case, and we can maintain the view that topic neutrality has bought us simplicity only by pretending that such work goes on, as it were, in another intellectual universe. It is better to take the approach I urge, for neither approach changes our actual mathematical practices but only our view of where nonstandard mathematics and logic are located; and the view I urge places it more in harmony with current scientific practice.

An observation: Recall from Part I, Section 11, that one objection to Quine's views that arises regularly in the literature is that Quine is forced to treat unapplied mathematics in a way that differs from applied mathematics, in terms of both how it is justified and how it is interpreted. We have found that the source for these troubles does not lie in a philosophically very deep place at all: It has turned out to be due only to the twin requirements that Occam's razor operate with respect to the ontological commitments of mathematics as a whole and that standard logic be topic neutral.[47] This is somewhat surprising, given how serious a problem for Quine's views this objection is taken to be. In any case, I can hope that because these principles are the source for such a popular complaint, philosophers will not have much trouble giving them up.

Let me, then, take stock. There are two tasks left to do before we can feel fairly safe that the picture being presented here is a viable one. First, there is the point, already raised, that the content of mathematical statements, and in particular, the ontological commitments of mathematical terms, cross systemic boundaries. For example, we recognize that any number of formal systems all pick out the standard model of Peano arithmetic. But I have yet to say how this sort of mathematical practice fits in with our talk of systems. Second, I

---

47 In particular, it does not seem to bear in any way on that aspect of Quine's approach that we favor: the view that mathematical objects are ultrathin posits.

THE STUFF OF MATHEMATICS

should deal explicitly with Quine's objections to the suggestion that mathematics and logic are conventionally true, because these objections, in some circles, are generally regarded as fatal to the view. I take up the second task first. Those readers who are already more or less convinced that Quine's problems with this view are not our problems should pass on to Section 6, where I turn to ontological matters, that is, to a discussion of how mathematical commitments cut across systems.

§5 Quine's Objections to Truth by Convention

Quine has several arguments against the view that mathematics and logic are true by convention. One argument purports to show that the notion of conventionality, as it is supposed to apply to logic and mathematics, and *not* to the empirical sciences, cannot be given any content. Another argument, one of his most famous, is supposed to show that the notion of truth by convention cannot apply to logic on pain of infinite regress. Yet another is designed to show that conventionality is, at best, a passing trait of sentences without lasting epistemic significance. In evaluating these arguments, my job is not to defend the positivistic position Quine directed them against but, rather, to show that they would misfire should a philosopher direct them against the view being developed here.

The first argument I will consider arises, to some extent, from Quine's holism. This is his claim that conventionality is a property that sentences have, at best, only in passing.[48] Certainly, a sentence can be taken to be true in a provisional and conventional manner as a postulate one adopts; but, eventually, should the sentence prove valuable to the conceptual scheme, its conventional initiation plays no further significant epistemological role. One may choose any set of sentences to practice drawing derivations from: These may be the ones originally conventionally chosen or some others; it doesn't matter.

Notice, however, that in Section 2 we did not take conventionality to be a trait of *sentences* at all but, rather, of *systems* of sentences. And the conventionality possessed by such systems is not due to the arbitrariness of choosing one particular postulate basis for a system rather than another but is an enduring trait of systems by virtue of their contrast to other systems. Focusing on the whole system this way rather than on particular sentences is nicely compatible with Quine's holism – up to a point.

Up to a point because we stop at the system rather than at larger units of the conceptual scheme (or the whole conceptual scheme it-

---

48 Quine 1954, pp. 119–20.

self). The main argument for regarding (pb)-systems as natural epistemic units arose in passing when we defined 'conceptual scheme' in Section 4: Although, when thinking in an idealized way, we may presume that our conceptual scheme can contain whole systems, in practice it can't. And, in saying this, I am not making a terminological point due only to the contrasting ways I defined 'system' (in Section 2) and 'conceptual scheme'. The claim is deeper than that.

To make things clear, let me leave aside momentarily those branches of mathematics and logic that are not applied to the empirical sciences. Part of the reason we practice (standard) mathematics and logic the way we do – that is, deriving such results within the context of systems rather than deriving them along with results from the empirical sciences without bothering to make a distinction in the sorts of things being shown – is that we only have access to logical and mathematical truths via theorem generating, and theorem generating is necessarily an ongoing practice. Should we derive a purely mathematical truth in a context where we do not keep track of what premises went into the derivation of which sentences, we will be unable to recognize the general applicability of the statement derived; we will, in fact, not know whether or not it is a statement that depends on (particular) empirical antecedents.

The foregoing is an epistemological point every bit as significant as is the fact that in doing empirical science one reaches for whatever sentences (taken to be true[49]) one needs, regardless of the academic borders drawn between disciplines. By taking conceptual schemes to be closed under their derivation relations, we idealize this point away and consequently overlook important differences between how we learn new facts about logic and mathematics and how we learn new facts about physics.

Notice that including nonstandard mathematics and logic in our conceptual scheme forces the system to become a natural epistemic unit, even if the foregoing considerations are not conclusive. This is because nonstandard mathematics and logic come packaged in systems, and we need to keep nonstandard mathematics distinct from the empirical sciences. But, as we've seen, systems are significant epistemically even apart from this last consideration, and I have stressed this because this way it is clear that we have an internal argument against Quine's view. For, on Quine's view, mathematics and logic enjoy a centrality in our conceptual scheme not possessed by sentences from the other sciences, and this alone forces one to keep track of the derivational ancestry of mathematical and logical truths.

Extending talk of conventionality beyond sentences to systems, however, does not escape a more important objection of Quine's. He

---

49 Notice that in this formulation I am still setting aside (for the moment) nonstandard logic and mathematics.

notes that one can design postulate systems for empirical subjects. One can take notions of physics, design postulates for them, and derive conclusions. In fact, to the extent that physics is theory driven, precisely this sort of thing goes on. Why not, then, conclude that physics, biology, or any other subject not too inchoate for axiomatic treatment is conventional in precisely the way logic and mathematics are taken to be conventional?[50]

The answer to this worry turns simply on the differences we have already marked between scientific practices in the empirical sciences and those in mathematics and, in particular, the justificatory practices in these disciplines. First note this. One way that mathematics has grown as a discipline is in the recognition that nearly any collection of concepts can be mathematized. By this I mean that one recognizes the presence of mathematics not by subject matter, not by the fact that what is being studied are numbers and shapes, say, but by the methodology employed. Consider one's favorite collection of physical concepts and an axiomatization of them. If one does not worry about whether the theorems generated are empirically realized, then one *is* doing mathematics and not empirical science. If, further, one studies various modifications of these concepts and various alternative axiomatizations, again without worrying about whether the theorems generated are empirically realized, one is doing mathematics.

By contrast, if one is doing empirical science, and not mathematics, one's theorems and the objects postulated by virtue of one's theorems take on additional epistemic burdens that mathematical theorems and the objects postulated by virtue of *them* are free of. First, we are supposed to establish epistemic connections between ourselves and many of these objects. Theoretical entities such as new particles may be postulated or derived from a theory, but then one goes on by trying to figure out some way of making (causal) contact. One's justification of a scientific theory, in fact, largely turns on how our subsequent attempts to *interact with* objects postulated by such a theory pan out.[51]

In addition, the justification of the *truth* of the theorems so derived is not merely that they have been validly derived, but in addition that they are *experimentally verified*, however that is to be understood in the particular science at issue. To put the matter in Quinean terms, what

50 Quine 1935, pp. 100–2.
51 To use the terminology given in Part I, Section 10 (and which will be revived in Section 6), mathematical posits are ultrathin and empirical posits are not. I should note that not every empirical posit carries the epistemic burden just mentioned – for example, *properties* don't. Let's distinguish between the property of gold and the gold items that have this property. Although there is an epistemic requirement that our knowledge of gold items relies (somehow) on the gold items themselves, I am unwilling to argue that our knowledge of the property gold relies on the property of gold. But this is a topic I cannot pursue further now.

is required is that the theory and its posits are successful in organizing our experiences, and this is something they can fail at. But the latter is simply not a requirement on mathematical results per se. And the *mark* of a purely mathematical practice is that this is *not* a requirement.

Here's an illustration. What made classical physics *physics* was that it could be (and was) overthrown on the basis of experimental evidence. Were it a mathematical subject, this would be irrelevant. Furthermore, there is an ongoing study that looks just like classical physics but *is* a mathematical subject. In it, for example, one studies classical point masses. And the results in that subject have not been overthrown by anything because they are simply not about the physical world. Classical point masses are mathematical objects in *precisely* the same sense that Turing machines are.

Here is another illustration. One can worry about the traditional classification of Euclidean geometry as a branch of mathematics rather than as a branch of physics. After all, it seems to have been seen as a study of *physical* space, and isn't that prima facie evidence that it is an empirical subject, a (failed) species in fact of *synthetic* a priori knowledge, as Kant would label it? My answer is "no," for Euclidean geometry was not studied in a way that allowed one to test the resulting theorems against experience in any way. One simply (perhaps) took it for granted that they applied. But it is the practice that marks out a science as mathematical rather than empirical, and this explains why Euclidean geometry is classified as a branch of mathematics rather than as a branch of physics, in contrast to ancient astronomy where the results were always checked against the movement of the planets. The subject matter of Euclidean geometry is two- and three-dimensional figures, and these are not physical objects. One does not test one's results by drawing actual figures and seeing if they obey the laws; rather, one uses such drawings as heuristics only.

Now one *may* worry at this point that I have erred too far in the other direction in my attempt to avoid the Quinean assimilation of mathematics to the empirical sciences; for one may think that I have included too much in mathematics. I am allowing just about any subject matter to be a mathematical one, provided only that one not worry about whether the results apply empirically. But in point of fact this is precisely what we have seen happen to mathematics in the last hundred years or so. For example, arguably, the topic of Turing machines looks like a study of *mechanisms* of a particular sort. What makes the subject mathematical, however, and *all* that makes the subject mathematical, is that one is not concerned with whether Turing machines can be physically realized. The fact that the mathematician is concerned with proof, and only with proof, when it comes to confirming his or her claims, has turned out to be the essential thing about mathematics that separates it from the other sciences.[52]

Notice the following, therefore. This objection of Quine's will seem to have force against the view I am developing only if one already has presupposed that the justification of mathematical results turns, at least in part, on the value the resulting mathematics has in application; for only in that case will it seem that the potential axiomatization of an empirical subject leaves us with no way to differentiate that subject from mathematics proper.

The last argument against truth by convention due to Quine that I will consider applies only to logic and not to all of mathematics: This is the claim that *logic*, in any case, cannot be true by convention.

I read the argument this way: First, a crucial (and undeniable) premise is that the sentences to be designated as logical truths form an infinite class.[53] This means at least this much: The conventions that are to fix the class in question cannot be ones that characterize the sentences once and for all, as it were, but only via steps. One must generate the sentences somehow.

At this stage, Quine poses a dilemma: We either use logic itself to generate these sentences from the conventions (in which case one is faced with an infinite regress), for

the difficulty is that if logic is to proceed *mediately* from conventions, logic is needed for inferring logic from the conventions. Alternatively, the difficulty which appears thus as a self-presupposition of doctrine can be framed as

---

52 The distinction drawn this way is a contemporary one, and therefore it does not always respect certain earlier views on where the border between empirical science and mathematics is located. Here is an illustration: Ancient astronomy was included in the mathematical sciences because its focus was on "saving the appearances." Ancient astronomers were not concerned with modeling the real motions of the celestial spheres or determining the causes of their motions, but only in predicting apparent motion. This made the subject a mathematical one in scientists' eyes in contrast to subjects where physical explanation played a role. One finds the same view, and same terminology arising, with regard to seventeenth- and eighteenth-century optics, which abstracted away from an actual description of how light propagated, and also with regard to Newton's theory. Contemporary critics regarded Newton's approach as *merely* mathematical because he gave quantified laws describing the effect of gravity without supplying an explanation of its operation in mechanistic terms.

From our point of view these are all examples not of mathematics but of empirical science because the focus is still on designing a set of principles that are tested against empirical phenomena. This simply does not arise as a possibility with mathematics proper. In fact one sees it in the perception of what the subject matter of mathematics is: geometrical objects that do not occur in a pure form in nature, numbers, functions of such, and operations on such functions. The metaphysical view of what such objects are contains, in a metaphysical code, a set of directions for how results about such objects are to be gotten, not, in any case, by experimentation.

53 This premise is undeniable simply because all the interesting cases involve logical systems which generate infinitely many logical truths.

turning upon a self-presupposition of primitives. It is supposed that the *if*-idiom, the *not*-idiom, the *every*-idiom, and so on, mean nothing to us initially, and that we adopt the conventions [needed to generate the logical truths] by way of circumscribing their meaning; and the difficulty is that communication of [these conventions] depends upon free use of those very idioms which we are attempting to circumscribe, and can succeed only if we are already conversant with the idioms. (Quine 1935, p. 104)

or such usage is *conventionally implicit* and the conventions are explicitly framed afterward. But the second suggestion is problematical because

[i]n dropping the attributes of deliberateness and explicitness from the notion of linguistic convention we risk depriving the latter of any explanatory force and reducing it to an idle label. (p. 106)

In evaluating this dilemma, it will be profitable first to consider a related case in epistemology where an infinite regress argument similar to the above plays a role. Alston has argued that there is a pervasive confusion among epistemologists between epistemic levels, that is, between believing A, believing that one believes A, and so on, being justified in believing that A, being justified in believing that one is justified in believing that A, and so on, and knowing that A, knowing that one knows that A, and so on.[54] Alston attempts to show how widespread this confusion is by examining it at work in a number of epistemologists. But I want only to consider his use of it to explain a skeptical argument of Sextus Empiricus. Here is Alston's gloss of Sextus Empiricus' argument:

In order for me to be justified in believing that p, my belief that p must satisfy the conditions laid down by some valid epistemic principle (for epistemic justification). But then I am justified in the original belief only if I am justified in supposing that there is a valid epistemic principle that does apply in that way to my present belief. And in order to be justified in that further belief there must be a valid epistemic principle that is satisfied in *that* case. And in order to be justified in supposing that . . . This series either doubles back on itself, in which case the justification is circular, or it stretches back infinitely. Thus it would appear that claims to justification give rise either to circularity or to an infinite regress. (1980, pp. 169–70, ellipses are Alston's).

Alston's moral is no surprise. As he puts it:

On the argument's own showing, what my *being* justified in believing that p depends on is the existence of a valid epistemic principle that applies to my belief that p. So long as there *is* such a principle, that belief *is* justified whether I know anything about the principle or not and whether or not I am justified in supposing that there is such a principle. What this latter justifi-

54 Alston 1980.

cation is required for is not my being justified in believing that p, but rather my being justified in the higher level belief that *I am justified in believing that p*. I can be justified in that higher level belief only if I am justified in supposing there to be a principle of the right sort. But it is only by a level confusion that one could suppose this latter justification to be required for my being justified in the original lower level belief. The regress never gets started. (p. 170)

Now, of course, to point out that one sort of infinite regress argument trades on a level confusion is hardly to show that every sort of infinite regress argument trades on a level confusion. In particular, although one's drawing a conclusion from premises on the basis of conventionally true inference rules can be taken to *justify* such an inference, it is not obvious (right off the bat anyway) that justification in this context is similar enough to justification in the context Alston is concerned with to allow us sanguinely to draw the conclusion that a level confusion is involved here too. Indeed, I don't want to claim that Quine is confused about the levels involved.[55] Rather, I want to show that he has saddled the truth by conventionalist with an overly strong requirement on truth by convention. Something weaker, which does not invite a regress, will do fine.

Quine, recall, focuses only on standard logic when he presents his regress; and there are two elements to standard logic in his sense. The first is that it is a particular sort of logic – "general logistic," as he calls it.[56] Let us generalize the case by considering any system at all, and let us ask what it means for someone to adopt such a system. Several points seem pertinent. First, as I have already noted, these systems can be characterized syntactically. That is, someone who adopts one can do so without knowing what the sentences in them are supposed to mean. Second, it would seem that all that is required for someone to be justified in asserting a sentence of a system is that they have derived it correctly using the inference methods given. Third, it doesn't *seem* to be a requirement on an individual having adopted such a system that he be able to show that he is correctly applying the rules.

Here is an analogy. Suppose I play a game of chess with my five-year-old daughter. Surely all that is required of her is that she move the pieces correctly. It is not required that she *show* that she is moving the pieces correctly. That is, in having moved the knight from one po-

---

55 On the contrary, *Quine* is under no illusions on this point. In Quine 1954, p. 115, he writes: "Briefly the point is that the logical truths, being infinite in number, must be given by general conventions rather than singly; and logic is needed then to begin with, *in the metatheory*, in order to apply the general conventions to individual cases." (Italics mine.)

56 This is one or another classical logic strong enough to serve as a foundation for classical mathematics: Quine made the distinction between first-order logic and what he came to call "set theory" subsequently. See Quine 1954.

sition to another, she is not required to open a chess book and show that it follows from the rules printed there that her move is a valid one. Let me put it this way: In order for her to be able to play chess, she must be able to *play* chess. In order to show that she *knows* how to play chess correctly, other skills are called for. She has to be able to show that the rules she is using are indeed the rules of chess, and that she has applied them correctly. It is easy to imagine that even if I have taught my five-year-old daughter to play chess she will be unable to do any of the latter tasks (although she may be quite capable of beating me at chess).[57]

Can we be sure that the same sort of shift in levels has taken place in the Quinean case? Yes, and this will be perfectly clear if we first consider a logical system that is very weak. Imagine that we are only employing the classical sentential calculus. Imagine, further, that the form of the calculus we use involves sentential formulas, the truth-table method for evaluating these formulas, substitution of interpreted sentences for these formulas, and modus ponens. In particular, imagine, as Quine does, that the convention of modus ponens is described thus:

M) *No matter what x may be, no matter what y may be, no matter what z may be, if x and z are true [statements] and z is the result of putting x for 'p' and y for 'q' in 'p $\Rightarrow$ q' then y is to be true.*[58]

Now suppose I infer 'John is running' from 'John is running and Sally is running'. The process, I imagine would go something like this:

(1) $(p \ \& \ q) \Rightarrow p$	tautology
(2) John is running & Sally is running	premise
(3) (John is running & Sally is running) $\Rightarrow$ John is running	subst, (1)
(4) John is running	M, (2), (3)

Now here is the crucial point. Following in Quine's footsteps to construct a regress, we need to take *(M) as a premise*. Our other premise is

(*) (2) and (3) are true and (3) is the result of putting (2) for 'p' and (4) for 'q' in 'p $\Rightarrow$ q'.

From these we are supposed to infer (4). Because this is an inference, it involves logic; and because it involves logic, it requires the

57 Indeed, if I purchase one of those chess-playing computers and play a game with it, it *will* beat me at chess. But it is utterly incapable of justifying that the rules it uses are valid.
58 Quine 1936, p. 103. I have taken some minor liberties in rewording the convention.

conventions by which logic is defined. At this point, Quine needs to invoke a convention regarding the every-idiom – for notice that idiom is used in stating (M) – before attempting to apply that convention in a derivation.

But we have to stop now to resolve a puzzle, namely, that somehow logical principles involving *quantifiers* have become involved here, although the language of quantification and the inference rules governing them are *not* part of the system we have adopted. How did this happen?

A clue is provided by the form of (M). (M) exemplifies the explicit use of semantic ascent. That is, we are no longer using modus ponens when we write (M) down as a premise; we are *talking about modus ponens.*

This also explains something else that might have seemed puzzling about Quine's discussion here. Recall Quine's claim (quoted earlier) that despite the truth by conventionalist's initial assumption that the logical idioms are meaningless, he finds himself presupposing those very meanings. Now we are in a position to understand how this happened: By indulging in semantic ascent, we saddle ourselves with a subject matter, namely, the very system we are studying, and consequently when we draw inferences about *it,* these inferences rely on *interpreted* logical idioms.

Let me add that it should be no surprise that attempting to write down the inference rules as premises drives the regress via semantic ascent. For consider the chess analogy again. If I know how to play chess, I know, as it were, what the legitimate moves are. Formulating these legitimate moves in a set of rules results in a set of rules that have as its subject matter what we might call the admissible states of a chess game. However, should I turn to the question of codifying the relationship of these rules to the game, that is, should I regard it as part of the game that I write down the rules and show that the moves are in accord with the rules, I will no longer be playing chess, and the subject matter of these new rules will be both the admissible states of a chess games and the rules about those states. Here it is perfectly obvious that ascent of Quine's sort changes the subject matter. This insight can be obscured in the case of general logistic by the fact that the same set of rules applies both in the system conventionally taken to be true and in the study of that system.

But perhaps we have been missing the point of Quine's argument here. After all, what is special about *logic* as opposed to other species of conventionally established sets of rules, such as those for games, is the fact that it is supposed to apply everywhere, including to topics such as itself. Perhaps it is this that forces the truth by conventionalist into Quine's regress.

Let's probe this possibility by considering a creation myth. Imagine that we attempt to establish logic by convention: We set up explicit rules and then decide that they will license inferences in every subject matter.[59] That is to say, the truth by conventionalist needs to stress the importance of distinguishing between the subject matter of logic and its application. This distinction stands even if logic is special among those subjects constituted by conventional rules in being applicable everywhere. We may formulate conventional rules that govern a game we call chess and then discover belatedly that instances of such games already exist *implicitly*. The same is supposed to hold of logic. Once the conventions are explicitly in place, I turn and reconstruct my logical practices everywhere, including in the metatheory of logic itself.[60]

One caveat: It doesn't seem to be a requirement on the doctrine of truth by convention that any logical system conventionally adopted be powerful enough to suffice as the logic for its own metatheory. Indeed, we have already seen a case of a logical system that isn't, and one cannot argue that adopting such a logical system by convention is thus ruled out. But let's consider, momentarily, only those logics that *are* strong enough to suffice as the logic for their own metatheory to show that adding such a constraint has no effect on the doctrine that logic is conventional.

Certainly, many (alternative) logical systems are powerful enough to function as their own metatheories. All that is required is that such a logic chosen have idioms powerful enough to describe the application of its own rules. This can be done, for example, by a first-order intuitionistic logic. Now, if we adopt first-order intuitionism, and we take Quine's line that logic is being presupposed in the framing of the conventions of logic, the logic so presupposed *can be* intuitionistic logic – and so with any other logic of this sort. At this point, it almost seems that Quine's problem is due only to his building into the term 'presuppose' a temporal element. To adopt a logic (make a choice), one cannot even get started without (previously) knowing what one is getting; that is, *in adopting a logic one must already have a metalogical description of what one is doing*. But it is hard to see what justifies a temporal interpretation of 'presuppose' here. I adopt a certain set of conventions. In doing so, I utilize that very set of conventions

---

59 Why not "they *and only they* will license . . ."? Simply because, in general, there is always a large number of inferences that are not deductive, given what deduction is taken to be at any particular time. For example, if the first-order predicate calculus licenses all and only logical inferences, then my inference from 'Jack is a bachelor' to 'Jack is unmarried' is not licensed by logic.

60 In this discussion, I have implicitly taken logic to be topic neutral. I write this way only because I am speaking, as it were, within Quine's idiom. See the considerations raised in Section 4.

adopted – something that can be *subsequently* recognized should I develop the sophistication to ascend to a metalanguage.

This may seem to raise the other horn of Quine's dilemma. Aren't we essentially dropping the attributes of deliberateness and explicitness from the notion of linguistic convention in that such conventions are only implicitly at work in the metalanguage? And isn't to do that, as Quine claims, to empty the term 'convention' of explanatory value?

We must first be perfectly clear about what is supposed to be *explained* here. For us, talk of conventions is not supposed to explain why this rather than that logical system has been applied (implicitly) to a certain subject area. The explanation for that turns on the value that the particular system has in being applied to that area: how simple it is, what impact it has on our practices there, and so on. 'Convention' only means this: There are many options to be chosen.[61] It does not need to explain anything else.

Perhaps, however, we have manipulated matters terminologically without really avoiding the problem Quine is concerned with. Alright, the Quinean might say, you've avoided my problem by making the issue of whether a logical system is true by convention a trivial one, but now the explanatory burden I am concerned with lies on the notion of 'application'. Let me, then, rephrase my question this way: In what sense can we say that a system has been *implicitly* applied to a subject area? As I did before in my complaint about 'convention', can't I complain that in dropping the attributes of deliberateness and explicitness from the notion of *application* we risk depriving the latter of any explanatory force and reducing it to an idle label?

By way of response, let us first consider an adding machine. What more can be asked for when, in describing this device, I show how the device is supposed to carry out the practice of addition (I show, in other words, what rules it has been designed to obey)? The machine does not utilize a metalanguage to describe the rules it obeys; rather, it simply obeys them. And I have provided an explanation for what it does when I state what particular algorithm it uses.

But people, one might argue, are different. One can't talk about what algorithm *they* have been designed to obey. Nor can one open them up to ferret out the dispositions they are relying on. So what is the claim that they are obeying certain conventions *implicitly* supposed to amount to here? There are two routes we can take.

The first is to exploit the fact that, in general, the statements that a community is willing to assent to are not present all at once: We can

---

61 Consider the conventions possible for obeying stoplights. What makes the matter one of convention is, as with the case of logic, the large number of options possible. What explains why this rather than that convention has been adopted is a matter of simplicity and other issues (e.g., the desire to minimize car accidents).

use the system we suspect to be implicitly at work in the community as a device for predicting what individuals in that community will and will not assent to (by way of inference) in the future. This approach has drawbacks. First, generally speaking, *many* systems are compatible with the test results because such results are invariably limited in number. That is to say, we can always consider alternatives that differ on inferences too large to be considered explicitly by the community. Second, one has the problem of separating out those results that are veridical from those that are based on errors of various sorts – what might be called performance errors. None of this is fatal to the enterprise, I would think. It simply makes things difficult.[62]

There is, however, a second option. This is to get the community to adopt an explicit set of conventions (of their choice) compatible with their practices. We then take the implicit conventions previously at work to be the ones that are now explicitly stated. Notice the following: Because the previous practices are finitary, there will certainly be slack between the explicit conventions and the practices (to date) they are supposed to codify. I am perfectly willing, in this case, to say that there is a certain amount of (philosophically insignificant) "indeterminacy" in the question of *which* conventions, precisely, are the ones that were adopted by the community implicitly; and, indeed, it is often perfectly clear that the logical inferences licensed by a community set of practices is both wider and narrower than the ones explicitly adopted.[63]

However this does not empty the system, taken to codify community-wide practices, of epistemic significance and explanation, for once the system is adopted, its epistemic significance is clear: We use one or another postulate basis to generate sentences, and the system has epistemic unity because of our need to keep such a practice intact. Furthermore, the explicit codification provides explanation even when it comes to the implicit logical practices in place earlier: Namely, such practices were not merely the adoption of a certain set of sentences, they were a set of inference practices, methods of deriving sentences from other sentences, and it is those practices we are codifying systematically.

Now I have been considering a more general problem than the one facing our truth by conventionalist. I have been concerned with the question of what sense can be made of the claim that an implicit set of

---

62 Indeed, these issues are identical to the sorts of problems that arise in the science of linguistics.

63 For example, involved in folk practices may be inferences that seem to rely on modalities, temporal idioms, and so on, none of which need be included in our explicit codification. Indeed, this could even be true of inferences based on quantificational idioms, as we have seen.

conventions is at work in the logical practices of a particular community. But the problem facing the truth by conventionalist is far more tractable than this problem. And the reason for this is that our truth by conventionalist doesn't have to worry about already existing community practices; he need only concern himself with how he is supposed to adopt a system from scratch. Thus *his* implicit adoption of conventions in the metalanguage is constrained by the system conventionally adopted for the object language. His implicit practice is fixed by a decision made when the system itself is conventionally adopted – and so all the way up the truth-theoretic hierarchy. Thus even if one thinks there are things to worry about in making explicit already existing implicit practices, such worries do not extend to the truth by conventionalist, whose implicit practices are explained by the fact that he can't simply adopt a logic by convention and explicitly apply it everywhere all at once.

I must briefly raise two other issues before bringing the discussion of this set of topics to a close. The first is that the reader may be disturbed by the infamous rule-following problem.[64] This problem is purported to arise regardless of whether the rules followed are implicit or explicit – and so it is a problem the lack of solution to which neither supports nor hinders someone who is concerned with answering *Quine's* dilemma. I confess that I, like everyone else in this business, have a story to tell about how to get around this problem, but it is not a story I want to tell now. For our purposes, it is best for the time being to treat this problem the way many epistemologists treat skepticism: It is a problem that must be dealt with *eventually* by any respectable philosopher working in the area, and *eventually* I promise to do just that (although not in *this* book).

The reader may have a second worry. Consider the alternative dilemma: Either (logical) reasoning is involved in applying the rules of logic (in which case one is faced with an infinite regress) or such usage involves brute dispositions (in which case the normativity of logic seems lost).

By 'the normativity of logic' I am not alluding to the "fact" that logical principles cannot be revised, for what can that mean in this context? It could only mean that the standard logical principles cannot be replaced, the logical system, that is, that we apply (to one or another domain) cannot be replaced, for no general restrictions hold on what properties systems can have in and of themselves. But we have not said anything so far about what restrictions exist for systems that are (potentially) applicable ones: Perhaps any logical principle in the

---

64 See Kripke 1982. The literature this has spawned is gigantic.

standard system can be replaced, and perhaps if a statement can't be, this is no more surprising than analogous facts that certain purely empirical statements can't be replaced either.[65]

What must be meant by 'normativity' here is simply that in applying the rules of a system, anyone (or anything) can make a mistake. But it is not obvious, when the rules are implicit, how one is to show that such a mistake *is a mistake* rather than a valid inference in the system implicitly adopted. My solution is implicit in what I have already written. When we adopt a set of conventions explicitly, we recognize deviations from it as "mistakes" in this pertinent sense of 'mistake'. Where the set of conventions is implicit, we rely on the subsequent explicit conventions that supplant the implicit conventions to define which earlier practices were "mistakes."[66]

Let me sum up. I see Quine's objections here as ultimately motivated by the attempt to replace one sort of justification of mathematics and logic with another. He desires to replace justification of mathematics and logic sui generis (i.e., via proof) with what justifies their application to certain subject matters. Certainly, if one is a holist, experiments in empirical science refute large units of scientific theory in which mathematics and logic are contained. But this is compatible with the view that what is open to refutation is not the truth of the mathematical and logical truths per se but only our license to apply them to the empirical domain in question. And it is this latter view that remains unscathed by the sorts of arguments we have examined here.

§6 Grades of Ontological Commitment[67]

We now turn to the question of how we must modify the picture of ontological commitment sketched in Section 2. In order to do this I first briefly recapitulate the various notions of ontological commitment that I have been developing in this book and then rehearse and

65 This *is* the line I will push eventually. See Part III, Section 5. I should also add that I am once again allowing talk that treats sentences as things that can carry their content successfully across systems.
66 Certainly this move is entirely benign when it comes to the truth by conventionalist, whose case we are primarily concerned with. But making mistakes is a subtle matter, and I have a bit more to say about it later. See Part III, Section 6.
67 Until now I have been working in a fairly general setting: the arbitrary system. But the discussion of ontology (especially the topics of Gödel's theorem and the standard model that follows) largely takes place against a classical background. Much of what I say applies outside the classical context, but I will not discuss the details of how much or how far in this book.
    I should also add that the framework of systems is itself studied from within the vantage point of standard mathematics. I say a little more about this toward the end of Part III, Section 3.

expand the nomenclature we have adopted to describe the possibilities available.

Let's say we're given a collection of existential claims about the extensions of a collection of predicates. If we understand these claims to be local, ones restricted only to a system Γ that is, and if there are no requirements on how we come to know about the objects posited except through the method of deriving theorems in Γ, then we call such posits *local* ultrathin posits. In this case, as we have noted, there is no sense to the idea that the terms in two different systems can refer to the same objects.

We may, however, intend such posited objects to be ones that *can be* referred to by terms in more than one system. If, however, there is nothing more to the requirements on how we come to know such posits than that it be by derivations of theorems in the context of such systems, we call these *nonlocal* ultrathin posits. We have not yet discussed such kinds of posits, but we will shortly, as I intend to argue that mathematical posits are nonlocal ultrathin posits.

Next we have *thin posits*. These also come in two flavors: local and nonlocal. But they differ significantly from ultrathin posits in that it is required of them that they contribute to the success of the empirical project of "organizing experience" in Quine's sense. That is, they have the added epistemic burden of having to be required in empirical scientific theories. They must, as Quine puts it, contribute to such a scientific theory's simplicity, familiarity, scope, fecundity, and success under testing. I will have little to say about such posits in this book, for two reasons. The first is that I suspect there are no local thin posits because of a point already raised (see Section 4): We tend to consider the ontological commitments of empirical science as a whole, and so it is invariably assumed that empirical posits are not restricted to the confines of a single theory. On the other hand, although I think there are many nonlocal thin posits (such as natural kinds and sets of empirical objects), these items are not central to the topics of this book, since I believe that mathematical posits, even those in applied mathematics, are ultrathin.[68]

Next we come to *thick posits*. As we have seen, these carry an additional epistemic burden carried by no posits so far canvassed: They must sustain (something like) a reliabilist epistemology, that is, not only must they possess the virtues in Quine's list in respect to our sci-

---

68 This is not to suggest that mathematical objects, when they arise in the context of applied mathematics, cannot possess the qualities of thin posits; for in the right circumstances, an ultrathin posit can function thinly. Indeed, it is a requirement on any branch of mathematics that, *if it is to be applicable to the empirical sciences*, its posits must possess the virtues of thin posits in regard to the field(s) the branch of mathematics is applied to. But I am not going to say a lot about this either because it does not directly bear on the range of topics I have chosen to write about in this book.

entific theories, but our knowledge of them must depend in a crucial way *on them*. Here is where causality, particularly the causal connections forged between speakers and what they talk about, must play a role, even if it is not the simple direct role given to it by causal theories of knowledge.[69]

It may seen that my sketch of the options available for types of posits leaves out a substantial possibility. Surely there can be posits in mathematics that have the virtues in Quine's list with respect to branches of mathematics, and *not* with respect to the empirical sciences. For example, consider the notion of free choice sequence as it arises in intuitionism.[70] The value of this notion, surely, is that it enables us to derive interesting results about the intuitionistic real numbers. Similarly, we find that the value of set theory, for the most part, is in its application to other branches of mathematics, as, to date, it has had no (direct) impact on the empirical sciences themselves.

The reason I am leaving this possibility out is that, in general, no mathematical object is *required* to pay its way, ontologically speaking, even as far as results in pure mathematics are concerned. I can define a sort of mathematical object and derive results about it without that sort of object, and results about it, having any application to any other branch of mathematics at all.[71] If the mathematics is sufficiently interesting, people will be drawn to show results in it anyway; otherwise the fledgling field will fall into oblivion. However, for better or worse, oblivion is not nonexistence. By contrast, a natural kind that does not pay *its* way (caloric fluid, for example) soon finds itself keeping company with the unicorns.

Now I have already given substantial arguments for why mathematical objects are ultrathin posits and nothing stronger. But the discussion of the history of mathematics (see Part I, Section 4) seems to make it clear that mathematical posits are nonlocal. How does this work?

Before addressing this question, it will be useful to introduce a couple of terms of art. Let us continue to describe the sentences that ap-

---

69 Although I have stipulated it in the text, it is not clear to me that thick posits must invariably possess the virtues contained in Quine's list. In particular, my knowledge of my acquaintances, famous people I have heard of, particular rocks in my garden, a comet I saw two years ago, and so on, if examples of knowledge, are so entirely because of reliabilist considerations and *not* because of how they contribute to successful empirical science (even if the latter is understood idiolectically). But I am letting the stipulation stand in the text because my concerns are not with thick posits except insofar as the requirements on them contrast with the requirements on mathematical posits.

70 See Troelstra 1969 or Dummett 1977.

71 Much, but not all, of number theory has this status. One can also mention in this context recursion-theoretic hierarchies, bounded arithmetic, modal logic, and (much) work in abstract algebra. And this is hardly an exhaustive list.

pear in systems as *sentences* when they are individuated in an entirely syntactic manner. However, let us call them *statements* if they are individuated not only in terms of their syntax but also in terms of what they refer to. Given this use of 'sentence' and 'statement', we will help ourselves to certain common idioms such as, 'sentence α *expresses* the same statement as sentence β'. Hitherto, our assumptions about systems have required that α expresses the same statement as β if and only if α is the same sentence as β, and α and β occur in the same system, but we are going to relax that restriction now.

Let's start, then, with a particular system Γ_1 in a language $\mathcal{L}_1$. Generally, if Γ_1 is at all interesting mathematically, it is syntactically incomplete, or at least, if it is unclear whether it is syntactically complete or not, it is hard to produce the proofs needed to decide certain sentences.[72] In certain cases, we can design a second system Γ_2 in a language $\mathcal{L}_2$ (properly containing the vocabulary of $\mathcal{L}_1$) that deductively contains Γ_1, but either decides sentences undecided in Γ_1, or is such that the proofs for the sentences we are interested in are easier to construct than the proofs in Γ_1.

On what grounds, however, are sentences appearing in Γ_1 taken to express the same statements as when they appear in Γ_2, and related to this question, what makes particular noun phrases occurring in the sentences of Γ_1 refer to the same things as when they appear in Γ_2? Simply this: *We stipulate it.*

Before going into what constraints on this sort of stipulation are available, we should sketch a slightly more complicated case that is possible (and not uncommon). This is one where we are willing to consider systems Γ_1 and Γ_2 as before, but where Γ_2 does not (quite) deductively contain Γ_1. Suppose, for example, that Γ_1 is syntactically inconsistent. In this case, we may be able to define a Γ_2 that contains the majority of the most interesting sentences that were derived from Γ_1 before its syntactic inconsistency was discovered. In such a case we are willing to regard the sentences the systems have in common as expressing the same statements.

In such cases, however, we cannot any longer take all the sentences appearing in both systems to be true. Those appearing in Γ_1 but not in Γ_2 we regard as *false*.

A few definitions are now in order. First, we say two systems Γ_1 and Γ_2 *monotonically overlap* if the two languages $\mathcal{L}_1$ and $\mathcal{L}_2$ that they be-

---

72 That syntactic incompleteness is going to kick in fairly early is due to the (mathematically speaking) weak requirements needed to derive Gödel's incompleteness results. That, in any case, proofs deciding certain mathematical sentences one way or the other are often hard to come by is a commonplace of mathematical practice. By the way, I am not suggesting that generating proofs is difficult – usually that is easy. What is difficult is generating the proofs for *certain* sentences that we'd like to know the status of.

long to are such that the vocabulary of the former is contained in the vocabulary of the latter. We also say that any finite linearly ordered set of monotonically overlapping systems have been *monotonically connected* if we resolve the potential truth-value differences among them in favor of the later systems. We describe the connection between such systems as an *inclusive* one if no sentence derivable from any system so connected is false, and finally, we call later systems (relative to the linear ordering) *descendants* of earlier systems.[73]

I now go on to give several examples of connected systems. The point of doing this is to show that, although it is theoretically possible to define connections among a large number of systems, in practice mathematicians are rather picky about which connections between systems they are willing to take seriously. We will consider what criteria there are for such acceptable connections after giving the examples.

(1) Consider first-order classical number theory and first-order intuitionistic number theory, and consider the obvious connection possible between them: the homophonic identification of the logical connectives and quantifiers and the homophonic identification of the extralogical vocabulary. Because a simple embedding is possible here, we do not have to regard any of the sentences appearing in either system as false. Rather, we can take the classical system to be one that allows us to derive much more about the objects in question than the intuitionistic system does.

(2) Consider ZFC and ZF + AD (Zermelo–Frankel set theory plus the full Axiom of Determinacy). We can connect these two systems in two possible ways, one in which the Axiom of Choice and those results depending on it are regarded as false and one in which the full Axiom of Determinacy and those results depending on it are false.

The examples I give of connections from now on are ones that mathematicians generally take seriously.

(3) Consider any system Γ_1 powerful enough for us to construct a Gödel sentence in the vocabulary of that system. We can connect a

---

73 Systems, of course, contain sentences, and not sentences paired with truth values. But because in our discussion of connected systems we are focusing on taking sentences to differ in truth values, more than sentences is involved. One way to flesh this out is to treat a connection between two systems as a mapping of the sets of sentences contained in the two systems into a collection of sentences paired with truth values. Rather than get explicitly technical about this matter, however, I will continue to speak informally and rely on the reader to supply details if they are needed.

I should add that this group of definitions can be broadened to handle situations where vocabulary drops out as we move from earlier systems to later systems; but as this changes nothing essential to the perspective I am arguing for, I omit the details.

Notice I have ignored the deductive relations involved in these systems because they are not the usual focus of mathematical practice. But clearly the above definitions could be framed (and generalized) with respect to these relations. I omit the details regarding this as well.

metasystem Γ_2 to Γ_1, where Γ_2 is a system containing a truth predicate for Γ_1, and is a system in which the Gödel sentence *is* derivable. This is an inclusive connection – any sentence derivable in Γ_1 is derivable in Γ_2.

Also available is a system Γ_3, which is in the same language as Γ_1 (no new vocabulary is involved in Γ_3) but contains as an additional axiom the Gödel sentence. This, too, is an inclusive connection.

(4) Consider any one of a number of standard embeddings of PA, and the language of PA, into ZFC, and the language of ZFC.

(5) Finally, I should mention a collection of examples that are vague simply because they predate the introduction of formal methods. Consider the various systems of numbers from antiquity until today. Such systems are collections of methods for generating number vocabulary and inference rules for generating sentences containing this vocabulary. Such systems supersede each other because the later systems possess stronger methods for deriving results and stronger methods for generating vocabulary. For the most part the connections between later systems and earlier systems are inclusive, though not always.[74] Consider, as examples of shifts from earlier systems to later systems, not only the introduction of new principles, such as induction, but the addition of new sorts of number vocabulary, such as negative number notation, complex number notation, integral notation, new methods of generating terms that are taken to be co-referential to older terms – for example, '$(2 + 3i)(2 - 3i)$' – and the more powerful inference rules that accompany the new vocabulary.

---

74 It is desirable to make a distinction here. On the one hand, there are cases where mathematicians make mistakes. It seems that something has been proved, and later mathematicians realize that the proof is fallacious. This is a case where mathematicians are temporarily mistaken about the properties of a particular system they are working in: They think a certain sentence is a theorem of it, and they are wrong. On the other hand, should we adopt ZF + AD, instead of ZFC, it might seem that no such mistake has been made. Rather, one system is being left behind for another.

Where it is not explicit which systems are involved, making this distinction requires some care, although by no means do I believe it cannot be made. For example, the use eighteenth-century mathematicians made of divergent series cannot be seen simply as a collection of mistakes. Such mathematicians were relatively clear that certain applications of these tools led to trouble and they avoided those applications. The nineteenth-century repudiation of such results was largely due to a desire for *rigor*, where use of this term indicated a desire for criteria – necessary and sufficient conditions – for the valid application of a method. Although later mathematicians saw their eighteenth-century predecessors as having been sloppy and prone to mistakes, it seems more correct to describe the less rigorous eighteenth-century mathematicians as using a different system with a large and open-ended number of ad hoc principles. Of course this is not to say that they didn't make lots of ordinary mistakes; of course they did.

See Part III, Section 6, where I take up this topic in a little more detail.

Now I have claimed that the identification of sentences and noun phrases across systems, and thus the identification of the objects referred to by the terms in these systems, are mere stipulations. And, theoretically speaking, it seems we have enormous, almost arbitrary, scope in what connections we may choose to make among systems. But as the examples above make clear, only certain connections are taken seriously.

We can best view these connections between systems as arising via a process where the later system is constructed from an earlier one it is connected to by changing its postulate basis in one of two ways. One may either add (or subtract) new premises or inference rules in the language of the system, or one may add new premises or inference rules that involve new vocabulary, or, more commonly, do both. Notice that the description of the descendants has been given entirely in syntactic terms; I have mentioned only vocabulary, premises, and inference rules.

I should digress momentarily to point out, however, that things are almost never perceived this way by mathematicians. Rather, their focus is almost always centered on ontology and epistemology. One starts with a body of methods, theorems, and axioms that are taken to be about a collection of objects used to prove results about these objects. And then the question posed is: How does one generate new methods for showing results about these objects?

Broadly speaking, two issues arise here. The first is this: Mathematicians often replace a system with a descendant in one of the ways just described, and we need to sketch what sorts of motivations they have for doing this. The second is that mathematicians often regard the descendant as containing terminology that refers to the same things that (some of) the terminology in the old system referred to, and we need to sketch what sorts of motivations are at work here too.

Let's take up first the question of what benefits a descendant should confer if mathematicians are to desire replacing a system with it. A short list follows.[75]

(1) First off, the shift in systems should be largely a conservative one, as mathematicians usually want to keep the interesting results. The ideal case is where the new system is inclusively connected to the old one; but where some of the proof techniques of the old system *are* troubled, this ideal must be dropped. It seems to me, though, that this ideal is *only* dropped when the proof techniques are troubled, where 'troubled' here means one of two things. Either, somewhat rarely, the system is out-and-out inconsistent or, more commonly, a particular

75 Notice the virtues forthcoming are all virtues involving sentences – syntactic objects. It is not being presupposed, that is, that sentences in the various systems refer to the same things, or, in other words, have content in common.

proof procedure is deeply mired in ad hoc qualifications and restrictions. Consider as examples the shift from naive set theory as it was practiced by Cantor to ZFC[76] and the shift from eighteenth-century analysis, especially in regard to divergent series, to the late nineteenth-century version.

(2) Such a shift should point mathematicians toward new results that can be shown in the old system. Consider as examples analytic number theory and the theory of functions of a complex variable. One is able to easily show a large number of results about arithmetic (e.g., arithmetic progressions) and about real functions (e.g., the integration and summation of real functions), respectively, using the stronger system, and many of these results can be shown directly, although often with more difficulty. Another contemporary example is nonstandard analysis: Many standard results become much easier to show using the additional power afforded by nonstandard analysis, and certain results, such as the Bernstein–Robinson theorem, were first shown using nonstandard analysis.[77]

(3) This is vague, and perhaps subjective, but such a shift in systems should make it easier to show results. Generally, the more proof-theoretic methods one has, the easier it is to show results, but there are factors also at work having to do with how easily mathematicians can utilize their capacity to visualize, and psychologically manipulate, images. I don't have much to say about this except to stress that it is really a factor. For example, it plays a role in explaining the appeal of nonstandard arithmetic. The vector representation of complex numbers and possible-worlds semantics for modal logic have similar virtues.

Another way it can be made easier to show results is if the shift in systems increases the surveyability of proofs. This is not necessarily just a matter of the proofs being shorter in length, although that is certainly a factor. Incidentally, this value is relevant not only to showing results but also in promulgating results through the profession. Quite a lot of work on the part of professional mathematicians involves streamlining proofs for results already shown.[78]

---

76 Moore (1982, p. 147) attributes to Zermelo the explicit desire to preserve as many classical theorems as possible.

77 See Davis 1977 for a delightful presentation of nonstandard analysis and, in particular, the Bernstein–Robinson theorem. Details about analytic number theory and complex variables may be found in any of a number of standard textbooks, for example, Cohn 1980, Nevanlinna and Paatero 1969, and Hille 1973. The point is so plain one can even see it at work in *exercises*. For example, see Spiegel 1964. Also see Maddy 1990, p. 145, where she makes the same observations about Martin's result, and the Axiom of Choice.

78 Clearly there is a value not only in deriving new results but in making it easier for professionals to understand the proofs of results already established. I discuss the significance of this further in Part III, Section 2.

(4) Such a shift in systems should enable mathematicians to generalize and unify mathematical results in the old system. There are at least two ways this can happen. First, one can find that what appeared to be a class of distinct results turns out to be (given the new system) instances of one particular phenomenon. Second, one can find that results that are fragmentary and suggestive in the old system are completed when one shifts to the new system. The various applications of complex numbers offer a beautiful example of this. Consider the impact of switching from the real field to the complex field on results from number theory, results about canonical forms for linear operators, polynomial equations, and finally, equivalences among exponential, logorithmic, trigonometric, and hyperbolic functions. Another example is the impact of the Axiom of Choice on transfinite arithmetic.

(5) Such a shift in the system should allow mathematicians to supply proofs one way or the other for open problems. Again, later, one may be able to show such results directly using the old system; but often this is *not* possible. Conservativeness is desirable here only so far as the results we already have in hand are concerned: If the augmented system, proof-theoretically speaking, goes beyond the system it augments, that is perfectly alright, provided the resulting system has a certain number of virtues. One thing one doesn't want: open problems resolved in an uninteresting way (say by out-and-out postulation), unless this is unavoidable. Again, these sorts of considerations have become quite explicit in set theory where one contrasts and compares alternative systems.

(6) Finally, the shift in systems should not involve *such* a gain in power that the new system is syntactically inconsistent. In the wake of Gödel's second incompleteness result, however, it seems one has only two ways of soothing such worries. First, one can supply relative consistency proofs.[79] More valuable, however, is the simple (although laborious) generation of results that hold in the system. If this does not turn up an inconsistency, one can tentatively, but only tentatively, assume consistency.

Now we take up the question of what convinces mathematicians that a change in the system used does not constitute a change in subject matter. Offhand, there may seem to be something puzzling to be-

I should add, however, that much of this streamlining goes on in textbooks rather than in journal articles, except in the context of papers which otherwise have new results.

79 The value of relative consistency proofs is a matter of some controversy, in part because under certain circumstances it is possible to design statements α such that (1) α expresses the consistency of an extension of PA, and (2) α is provable in PA. See Feferman 1960 for the result, and Resnik 1974 and Detlefsen 1979, among others, for discussions of its philosophical implications.

gin with about the admissibility of treating a descendant of a system as having terms that refer to the same things that terms in the older system refer to. For on the view I have been pushing, it does seem that a mathematical topic is defined by virtue of the system given. If one wants to study a particular sort of object, one simply stipulates the properties desired. Provided that the result is not boring because the stipulations have inadvertently generated a syntactically inconsistent system, we're done (or so it would seem). For once one has a particular system in hand, if the result of expanding it doesn't lead to redundancy (because the new results can be derived in the original system), the expansion seems open to the objection that we have changed the subject.

Indeed, in many cases, this is exactly how matters are seen. For example, consider the axioms for a group. If one augments these axioms with a requirement that the group operation be commutative, one now has a set of axioms for *abelian groups*. Abelian group theory is not considered to be the same subject as group theory per se, although the subjects are closely related. The difference is that not everything that holds of abelian groups is taken to hold of groups in general. Our question is simply why one doesn't always see a change of subject whenever one sees a change of system.

Historically speaking, a number of factors have been at work, and we go on to mention a few of them.[80] I stress that when looking at historical cases, one will usually find several of these factors working at once.

First is the force that various methodological and extra-mathematical background beliefs exert on the direction of mathematical research. For example, consider one or another version of the view that the objects that mathematical statements are about are not located in the world around us. If this view is accompanied with an implicit belief that such mathematical objects are elegant and simple in certain ways, and an implicit epistemological view that access to them is through some sort of intellectual intuition (perhaps of the sort Des-

80 Two caveats: First, there is a bit of rational reconstruction going on here. Because mathematicians generally do not view matters in terms of systems and whether certain connections between such systems are desirable, but instead focus on ontological and epistemological questions, they are prone to ask not whether a system Γ_2 should be connected to a system Γ_1 but, rather, whether the new vocabulary in Γ_2 refers to anything "real," and whether what is derivable via Γ_2 is "true." But this use of the idioms 'true' and 'real' hides what I take to be *really* going on.

Second, it seems to me that as mathematicians have become more familiar with formal systems and their properties, the tendency to see a change in subject matter whenever the system is changed has intensified. So, with the one exception involving Gödel sentences, the factors discussed immediately below are not so powerful in their impact on mathematicians as they were during the nineteenth century and before.

cartes thought we had), it allows the mathematician to shift on the system without seeing a change in topic, for she can console herself that she simply got certain things *wrong* about the objects involved.

Another example is the view, widespread in the seventeenth and eighteenth centuries, that the *physical* universe has mathematical structure built into it. One finds this sort of view expressed regularly in the writings of Galileo, Kepler, Descartes, Pascal, Newton, and others. Because the subject matter of mathematics is taken to be embodied in the world in a certain way, such a view also allows one to shift on the systems one is operating with without thinking that the subject matter has changed.[81]

A second factor is the desire to show results. Mathematicians are willing, almost subliminally, to shift on their definitions, if in doing so cases are excluded that prevent general theorems from being established. Sometimes the results wanted come first, and the definitions of the objects that have the properties shown come second. Part of what supports the perception that there has been no change in subject matter in this case is the fact that the theorem desired dictates the definition given, and part is that the alternative earlier discarded didn't seem to amount to much in the first place.

There are many illustrations,[82] but the history of the shift from naive set theory to (one or another) formal set theory illustrates this phenomenon fairly well. Russell, in developing and propounding his theory of types, did not think that he was changing the subject matter of set theory. In this case, interestingly, what was desired was the presence of certain results and the absence of others.

Another example is the development of set theory in the hands of Zermelo. Moore (1980) argues that a primary motivation of Zermelo's axiomatization was to provide a defense of his proof of the well-ordering theorem. Indeed, what is striking is that neither Zermelo nor anyone else participating in the heated debate about the Axiom of Choice thought there had been a change in subject matter given Zermelo's explicit axiomatization. (Indeed, around the same time, Cantor's own unpublished proof of the well-ordering theorem was

---

81 Why, then, do I take the subjects under study here to be branches of *mathematics* rather than, say, physics? The answer is that, although seventeenth- and eighteenth-century mathematicians lost sight of the distinction between physical arguments and mathematical arguments to such an extent that they allowed themselves to use physical arguments in proofs in a way that Archimedes over a thousand years before would not, they never lost sight of the fact that numbers, functions, derivatives, and so on, are not physical objects. What justified, in their eyes, the use of physical analogies to help derive purely mathematical results about such things was the fact that the physical universe was taken to have a mathematical structure – not that somehow the proper objects of mathematics (numbers and functions, for example) were somehow physical.

82 A detailed classical discussion of this phenomenon may be found in Lakatos 1976a.

circulating. No one suggested the two men were working in different fields.[83])

A third factor is this: The subject matter under development is often constrained by particular applications that the developers have in mind. These applications can be applications either outside mathematics proper or to other branches of mathematics. In such cases, mathematicians will allow shifts from one system to another (without regarding the subject matter as having changed), provided that they take the objects they are trying to study to be ones that are ideally suited for the presupposed applications. The project is one of trying to formulate a definition of the type of object that will best apply, and one doesn't always quite get it right at first.[84]

Again, consider the development of the theory of types in the hands of Russell. Here we find much shifting in the systems involved without a corresponding perception of a change in subject matter. A primary factor allowing this, it seems to me, was the constraint provided by the project of supplying the mathematics of Russell's time with a foundation. This meant that whatever replaced naive set theory had to be powerful enough to do what naive set theory could do, and focus on this constraint provided an impression of continuity of subject matter for Russell and his immediate successors (e.g., Ramsey).

A fourth factor, especially in the early history of mathematics, is simply a failure to recognize what we, with the benefit of hindsight, see as the other possibilities available at the time.[85]

Consider elementary arithmetic. The background field of applications for counting does little toward explaining the acceptance of the principle of induction by the ancient Greeks, as adopting such a principle had little effect on the applied mathematics of that time. What is also surprising is that the principle seems to have been adopted with no controversy. Perhaps this could be explained by the presence of an overwhelming (second-order) intuition that the *numbers* are closed under a certain (idealized) method of generation. But this sort of explanation should be avoided because it relies on the numbers themselves to explain the introduction of induction. Better to point

83 For a masterful presentation of the history here, see Moore 1982.
84 I don't want to give the impression that the implicit field of application for a branch of mathematics must remain the same during the development of that branch. On the contrary, there can be shifts of all sorts in the presumed area of application, including ones where the area shrinks. For an example of this in probability theory, see Daston 1988. Naturally enough, shifts in the purported field of application can change the constraints operating on a particular system, and this in turn can provide motivation for changing the system, again, without a perceived change in subject matter.
85 Putnam discusses this sort of phenomenon at length in his 1962.

out (what seems obvious anyway) that the other possibilities, ones that we recognize with hindsight as available, simply did not occur to anyone at the time. Nobody had to argue that *this* is the way numbers are because nobody saw that there was any other way for numbers to be.

This sort of factor often provides the explanation for those cases where shifts from weaker systems to stronger ones are unaccompanied by controversy of any sort: If no one regards the shift as a change in subject matter, it is because the failure to see the other available possibilities gives the impression that the new inference rules or axioms *obviously* hold of the objects in question.

Euclidean geometry provides a nice example of this same sort of failure. As I've already mentioned, the ancient discomfort with the parallel postulate was not a fear that a controversial axiom was being adopted in place of other alternatives. Rather, the worry was that the axiom was too complicated to be an axiom. But clearly mathematicians thought it obvious enough to attempt, for centuries, to prove it from the other postulates.[86]

Finally, there is the fact that mathematical practice almost always takes place in an informal medium – namely, one or another natural language – which often makes it very difficult to determine the outlines of the system being used to derive results. As a result it is simply much easier to shift on the system being used without any awareness that such a shift is taking place. Also, it is easy for controversies to arise between mathematicians because the systems they are using are different, although the informal medium they are working in hides these differences.

All the factors so far considered help give mathematicians the impression that there has been no change in subject matter, whether the change in system involves the introduction of new vocabulary into a system or only the introduction of new axioms or inference rules. I should stress again that the focus, interestingly enough, is never on the old vocabulary and whether it refers to what it referred to previously – that is taken for granted. Rather, the focus is always on the new vocabulary. It has to be shown that the new terms are compatible

---

86 In note 80 I observed that because contemporary mathematicians are much more aware of when such shifts are taking place, it can happen that such shifts will be much more recognizable as changes in subject matter. But not necessarily. Consider the topic of bounded arithmetic (see Parikh 1971 and Wilkie and Paris 1987, among others). Mathematicians (or certain logicians anyway) are interested in analyzing the proof-theoretic properties of number systems weaker than PA. However, it does not seem to me that this topic is regarded as a change in subject matter, except insofar as semantic ascent is involved. The *perception of the ontology involved*, that is, has not shifted. It is still *numbers* that are being studied, though one is studying how one's proof-theoretic tools affect what one can show about these objects. But I could be wrong about the perceptions of the workers in this area.

with the old vocabulary in certain ways. For example, if the terms are taken to refer to numbers, then it has to be shown that such numbers have the right properties – that is, that enough of the standard operations and relations are definable on them. Thus one thing that hindered both the introduction of negative numbers and complex numbers was problems with how such numbers should be ordered relative to the positive numbers. Ordering problems also plagued the introduction of transfinite numbers.

The appearance of continuity of subject matter, despite changes in the system that constitutes that subject matter, has been found to some extent to be a magician's sleight of hand. This should be no surprise, for we have allowed no space for the possibility that the objects under study themselves exert a force on the direction research takes. Instead we have found the factors at work here to be ones having to do with application, the desire to prove results, the simple failure to recognize alternative possibilities, and broader background beliefs of a methodological or philosophical nature.

Now it does seem that all of the factors mentioned so far, though explaining why mathematicians would be willing to think there has been no change in subject matter, hardly *compel* taking such a viewpoint. In fact, it may seem that there would be no effect on mathematical practice if mathematicians were to adopt the attitude that in all such cases there is a change in subject matter that mathematicians (historically) failed to see because of these factors. So, for example, one would say *not* that Weierstrass had finally placed analysis on a secure foundation but rather, that he had replaced a previous subject with a new one that was superior in several respects. And similar remarks could be made about every other shift in mathematical practice. We might even go so far as to argue that there is nothing wrong with informally writing and speaking as if there were no shift in subject matter in mathematics when these changes occur, simply because of the convenience of keeping the same terminology. But we would be careful to note that as philosophers we should realize the true state of affairs – that, ontologically speaking, there is nothing here that *requires* anyone to claim that, as I put it earlier, mathematical posits are nonlocal.

Let me summarize where we have gotten to. We have explored how the objects referred to by terms in different systems can be identified as the same, and we have recognized several sorts of reasons that have historically played a role in coaxing mathematicians to so regard mathematical objects. Before actually evaluating the desirability of regarding mathematical objects as transsystemic in this way, I want to consider yet another way that mathematical objects seem to be treated as independent of systems. After doing so, I will examine Gödel's results as a sort of case study in attitudes about mathematical ontology, and then draw conclusions.

§7 Multiply Interpreting Systems

We have explained how two different systems can have terms in their sentences that refer to the same objects. But mathematicians often seem to *distinguish* the sorts of objects that the same system can be about. Group theory is a fine example. *Any* sort of mathematical object that obeys the group axioms is a group.

Number theory can be regarded in the same way. We can take any sort of mathematical object that obeys the axioms of PA as a, say, "number collection." In the case of group theory, the axioms were formulated deliberately to apply to many different sorts of mathematical object. This is not true of the number-theoretic axioms – they were formulated with a particular sort of mathematical object in mind, and later it was realized that the terms appearing in the (first-order) theory can be reinterpreted. But the distinction here seems to be an accident of history. There is nothing in principle that prevents one from treating the first-order number-theoretic axioms the way the group axioms are treated.[87]

It may also seem that our capacity to reinterpret theories conflicts with the view that mathematical posits are local, that they cannot be referred to from the vantage point of more than one system. As it turns out, however, this practice does not necessarily drive us to be so ontologically ambitious. For consider: Reinterpretation of a theory takes place against a background theory. One recognizes the presence of a group operation among mathematical objects that have been picked out in some other way, for example, a certain class of functions.

This suggests the following account of multiply interpreted theories. They are not *systems* in *our* sense at all. When one recognizes that a group is present, this is only the recognition that a certain pattern of sentences occurs in a system. We do not, that is, have to regard a proof in group theory as a proof in any particular system but, rather, only as a schematic derivation of *sentences* the pattern of which appears in many different systems. Group theory, on this view, does not, strictly speaking, have a subject matter; it is, rather, a process of schematically marking out proof patterns among the members of an indeterminate class of systems.

This view is not so farfetched as it may sound. In particular, it does not imply that there are no groups; for let us revert to the analogy

87 In fact, they sometimes are treated this way. See Kaye 1991.

with games that was explored in Section 2. Consider the pawn in chess. This piece moves according to certain rules. Now it is easy to imagine that different board games could be designed in which there were pieces obeying the same rules pawns do. In such a case, one could characterize certain games as being ones in which pawns appear, and one might even be able to show that any such game would have certain properties. This would not imply, of course, that there is a game called Pawn which only involves pawns and which is being played when one marks out what properties any game with pawns in it has. That is, it is compatible with the concept of local thin posits that one can characterize and describe general properties of posits from outside the system without in fact actually referring to them in a way that violates nonlocalness.

When it comes to number theory, therefore, we should distinguish two cases. The first is where we treat PA as a system. In this case, the terms refer. The second is where we study a class of sentences homophonically identical to those of PA. In this case, our terms do not refer to anything. Rather we are studying a schematic set of sentence patterns that may occur in any number of determinate systems.

Similarly, when any system suffers reinterpretation, the view urged here is that the sentences of that system are actually being treated schematically; and, strictly speaking, to say that the reinterpreted set of sentences has a new subject matter is something that should be said only within the context of scarequotes.

The suggestions just made resemble in some ways those put forth by certain structuralists.[88] But here are the crucial differences. First, the study of such patterns of sentences presupposes an ontology of mathematical objects that sentences exemplifying such patterns hold of. We are not contemplating a position that replaces systems across the board with the study of these structures. For one thing, it does not seem true to mathematical practice, which does distinguish between group theory, on the one hand, and theories that are taken to refer to a particular class of objects, for example, number theory (as it is traditionally studied), on the other.

Second, if one is a set-theoretic foundationalist, one wishes to limit one's ontological commitments in mathematics to set-theoretic ones (of a particular set theory). In this case, it is tempting to replace the view that PA is a system that we have embedded in one way or another in set theory with a structuralist view of PA. We are not so tempted because we have not found much reason for adopting set-theoretic foundationalism.[89]

88 See the citations given in Part I, note 17.
89 I discuss this possibility a little more in Section 8.

Third, we have gone structuralist not over the objects referred to by sentences in systems but over the systems themselves. Our patterns are not ontological, but sentential.

Let me sum up. We have managed to preserve two suggestions presented in Section 2, that systems that are deductively equivalent should be identified and that mathematical posits are local ones, in the face of the common practice of reinterpretating axiom systems. The solution is to distinguish when an axiom system is being treated as a *system* in our technical sense, and therefore carrying ontological freight, from when it is being treated schematically, as representing one or another pattern of sentences in one or another system. Now this view is not one I am going to stay with, because I am eventually going to accept the view that mathematical posits are nonlocal, and therefore that the same posits can be referred to from within the context of more than one system. In this case, therefore, we will distinguish systems that are deductively equivalent when we understand them to be about different objects. Since this is the view that is eventually emerging, why the contortions for the last eight or so paragraphs? The reason is this: I am attempting to determine exactly what the factors are that prevent us from treating mathematical posits as local ones. And what has been made clear so far is that the widespread practice of reinterpreting axiomatic theories, contrary to first impressions, does not require nonlocalness.

Let us now turn to Gödel's results. Prima facie, they seem to offer a method of generating an open-ended number of new axioms for those systems to which Gödelian methods can be applied. What we are going to explore now is the question of what presuppositions are built into this view of these methods. The primary question I will try to answer is whether Gödel's results provide a motivation for treating mathematical posits as nonlocal.

Here is the first claim: All that the *proof* of Gödel's theorem shows, when it can be applied to a system, is the *syntactic incompleteness of that system:* One has constructed a sentence such that neither it nor its negation can be derived from the system in question. This means that either the sentence itself or its negation may be added successfully to the system without syntactic inconsistency being the result. Nothing requiring the suggestion that more than one system can refer to the same mathematical objects is involved yet.

The considerations lately raised about reinterpreting systems show this. For consider, as an example, a reinterpretation Γ^* that is deductively equivalent to a consistent axiomatizable extension Γ of Q (Robinson arithmetic), but has as its "subject matter" (note the scarequotes) a mixture of "numbers" and the syntactic strings of Γ^*. That is, given a particular Gödel numbering, let a numeral n refer to the string of Γ^* it is the Gödel number of, if it is a Gödel number, and let it refer to the "number" n otherwise.[90] Now, Gödel's results show that Γ^* is syntactically incomplete. But because Γ^* is deductively equivalent to Γ, we have shown that Γ is syntactically incomplete too.

However, consider PA again. If the Gödelian sentence is added to PA, the resulting system is still taken to have as its subject matter the standard model. On the other hand, this is taken to be false if the negation of the Gödel sentence is added to the system.

What is the source of this difference in attitude? It must come from an implicit understanding of another system: the metasystem (containing a truth predicate) in which the Gödel sentence *is* derivable.[91]

The presence of metasystems seems to offer a powerful intuitive push for treating mathematical posits as nonlocal. But what is actually involved in a system being a *meta*system for another system? Two things at least: Each term of the object system that refers has associated with it a new term in the metasystem that is *stipulated* to be co-referential with it, and second, the metasystem must have the power to describe the object system itself.[92] But notice that as soon as we allow this, we are accepting the nonlocality of mathematical posits. For we are allowing that terms in some other system can co-refer with the terms in this one. This shows that the intuitive plausibility of treating a metasystem as a metasystem *presupposes* taking mathematical posits to be nonlocal.[93] It has turned out, therefore, that Gödel's technical results do not require the view that mathematical posits are nonlocal. If one interprets his results so that where they apply they involve the construction of sentences that are true but not provable, one is already presupposing nonlocalness.[94]

With this last example in mind, let us return to the question of what value there is in treating mathematical posits as nonlocal. I offer a short list.

(1) Generally speaking, continuity of terminology is always appealing, if one can get it. Provided that no troubles arise by treating ear-

---

90 Notice that the term 'number' has scarequotes too. This is because, by assumption, if Γ^* is different from PA, it cannot have numbers as its subject matter as it is a different system, PA, which has terms that refer to *them*. Notice, also, that I assume Q and PA are consistent.

91 The truth predicate must occur in the metasystem in such a way that it is not conservative (relative to the object system). See Parsons 1974a and 1974b.

92 Of course this need not be done in the Tarskian way I describe here. For example, one might simply add substitutional quantifiers to the system itself, as I illustrate in the appendix. Doing so does not affect the issue under discussion, for it is still possible for one to think that no terms in the resulting system co-refer with any terms of the original system.

93 At least one mathematician I know of questions the nonlocality of mathematical posits, although he does not use this terminology. See Isles 1992 and Isles, unpublished. Actually, his full position is more radical, for he allows the possibility of the reference of a term changing within the course of a particular derivation and, therefore, presumably, within a particular system.

94 This reverses the position I took in Azzouni 1990.

lier terminology as referring to the same things as later terminology, one minimizes what would otherwise be a terminological maelstrom. With mathematics, treating cases where one has continuity of terminology as simultaneously cases of continuity of subject matter is especially harmless because the posits are ultrathin – one incurs no further epistemic and metaphysical obligations.

(2) Related to this is the psychological ease introduced by treating mathematical results as about *objects* of a certain sort. Object-centered thinking is intuitively easy because of its widespread presence outside mathematical contexts. Thus it is easier to treat the various systems that are taken to be about numbers as *about* numbers and in that fashion to tie them together. Clearly a crucial element in the traditional platonistic approach is thinking of the subject of mathematics as about *objects* of a certain sort. Such a view would not be so widespread and natural among mathematicians if it did not contribute to their ability to do what is important: prove results.

(3) Because (many) mathematical terms are applied terms, we need to be able to combine (some of) our mathematical results with those in the empirical sciences and everyday life. But this requires that we allow mathematical terms to apply freely in contexts outside their own systems. Of course we could always use new mathematical terms in such contexts, restate the rules of the postulate basis, and derive the needed results on the spot, but this would be fatiguing.

(4) Similarly, it is valuable, when one finds echoes of a certain mathematical structure appearing in other mathematical structures, to be able to import the same terminology and results with a minimum of anxiety.

In short, the group of factors motivating the treating of mathematical posits as nonlocal is two-pronged. On the one hand, we have the considerations of pragmatic convenience arising out of our psychology, our desire to apply (some of) our mathematics, and our desire for terminological continuity. On the other, there is the fact that taking mathematical terms to be co-referential across systems this way does not incur any new epistemic or metaphysical burdens; in short, there is absolutely no harm in doing so.

Let me sum up. We have investigated what factors drive mathematicians to treat mathematical posits as nonlocal – to require, that is, terms from different systems to be co-referential. These factors have turned out to be only pragmatic ones, largely involving terminological and psychological convenience. There is no harm in indulging in such desires provided that the result does not lead to confusion in practice. Nonlocality of ontological commitment, however, does not lead to mathematical confusion, for just as mathematicians may study any system they choose, they may also connect any systems they choose to, and it is hard to see what *mathematical* problems such practices could lead to.[95]

The second sort of confusion possible is philosophical. Philosophers can become troubled about the ontology and epistemology of mathematics, and I think it is clear that the sorts of ontological commitments mathematicians have been willing to adopt *have* bred such confusions. In particular, worries arise about how different mathematical terms manage to refer to the same things, and how mathematicians know what they know about such objects – in short, the sorts of worries arise that this book is an attempt to deal with. And here what needs to be done is exactly what we have been attempting to do: to sort out exactly what mathematicians do, and, as a result, show that these sorts of philosophical problems are not genuine.

One last observation. The reader may be worried that the picture I have given here makes mathematics *too easy.* One merely sets up a system and derives results. One can also, if the inclination strikes, arbitrarily connect systems with other systems to make one's job easier (as if it weren't easy enough already). But we *know* that mathematics is *hard* and that it calls for a high degree of creativity. Is this explicable on the picture I've given?

Well, yes it is. First notice that working *within* an algorithmic system is not merely a matter of mechanically cranking out derivations. It *could* be if we were obstinate enough, had the time and energy, and otherwise were rather unimaginative. What is so fatiguing about this approach, though, is that *almost all* the results we would generate this way would be boring and uninteresting: They would, in fact, be obvious variants of stuff we already knew. In practice, one produces proofs in *somewhat* the same way that humans play chess. We don't search out all the possibilities exhaustively; rather, we see patterns, proof patterns, that we then sketch out informally. Seeing these things, this way, is not a mechanical matter. In other words (peculiar as it may sound), we do not work within algorithmic systems *algorithmically,* or at least, we do not do so for the most part.[96]

Next, it is worth pointing out that there are hard facts about which systems can be connected to which *if* we are unwilling to give up results. Quine has stressed conservation as a virtue to be weighted with others when considering changes in one's conceptual scheme; but the mathematician has elevated conservation in this sense to an obses-

95 I discuss this issue a little further in Section 8 by focusing on the particular example of the so-called standard model of PA.

96 This fact is put to philosophical use later. See Part III, Section 4.
   Notice also that this fact enables us to avoid a standard objection that has been raised against traditional deductivism (see Resnik 1980, pp. 131–2), namely that, "throughout the history of mathematics mathematicians have been convinced completely of results by means of arguments that fall short of deductive proof."

sion. Where application is at issue, one can think that conservation is important because one likes to retain something if it works. But mathematicians treasure it regardless of whether the mathematics is applied or not. As I said, unless the system is troubled – inconsistent, for example – one wants to retain all of one's previous results. Furthermore, when augmenting the results of a system by embedding it in another, one wants to get new results not in a piecemeal fashion but dramatically, all at once as it were, and by principles that are broad-ranging and simple. Systems that augment other systems in this way are rare, and it is not obvious, to understate the problem greatly, how one is to construct them.

These values make the project of augmenting systems an arduous one. For example, consider Riemann integration. This is an operation that applies to a very large class of functions. It is based on rather simple principles; however, there are many functions to which it cannot be applied, despite the fact that we have an intuitive sense of what their integrals would look like if they existed.[97] If we want to extend the notion of integration to a larger class of functions, we want to do so in a way that respects Riemann integration; that is, we want an operation that gives us the same answer as Riemann integration does on the functions for which Riemann integration is defined. Constructing a system that augments Riemann integration this way is a nontrivial problem; and there is no reason a priori that it can be done at all, let alone in more than one way.[98]

One can ask what the source of these mathematical values is. The traditional platonist may feel that she has an explanation: The mathematician is studying a particular group of objects, and to change the laws governing these objects is to change the group of objects being studied. But I think this is a bad explanation, for two reasons. The first is that a weaker explanation will do the job of providing a rationale for why mathematicians have these values, and this is the (false) impression on the part of mathematicians that they are studying a group of objects. In general, realism is never needed to explain why a group of professional scientists have the practices they have: The collective delusion that what they are studying exists can do just as well. This is not to say, of course, that realism might not be needed to explain the *success* of such practices (but this is another matter).

The second reason I don't like this explanation is that realism is perfectly compatible with thinking that one was wrong about the ob-

97 The textbook example is the function defined on the open interval (0, 1), which takes the value 1 on irrational numbers and 0 on rationals. Intuitively, the value of the integral of this function from 0 to 1 should be 1 (the function is 1 almost everywhere in its domain). But the Riemann integral cannot be defined for this function. See any textbook on Lebesque integration, for example, Hewitt and Stromberg 1965.

98 As things turns out, of course, it can be done. See Hewitt and Stromberg 1965.

jects one is studying (and therefore it allows us to consider changing the laws that we take to apply to these objects).[99] And mathematicians, as we've seen, are perfectly willing to augment the laws that apply to a collection of objects, provided doing so yields new results without disturbing old ones.

What kind of explanation for these values *is* available here? Let us momentarily consider games again. Notice that when playing a game, it is not appropriate to change the rules. If one does so, one is playing a different game (or cheating, if one tries to do so during a game and foist the change on others). If we ask why this is true of games, one or another psychological or sociological explanation will be given. Certainly an important factor is that games are often competitive, and one has taken up a challenge in taking up a game. Changing the rules can eliminate this challenge.

I think a similar explanation is called for in the case of mathematics. The profession is engaged collectively in problem-solving behavior, and problems are not solved (usually) by changing the ground rules that gave rise to them. I think this largely explains the value of conservation in mathematics. What one needs to ask is why it is allowable to change the rules as much as it is, that is, why it is such common practice to augment the ground rules.

I think the answer is that doing so in a manner that is perceived as an acceptable solution to a problem (or set of problems) is as hard or harder than trying to do so directly. Furthermore, the result almost invariably creates a domain of new problems and structures that are interesting and intricate. Mathematics is a more creative and intriguing subject than anything afforded by a game, and this helps explain why changing the ground rules is acceptable: The new "game" that results may simply be of more interest, in many ways, than the old one was.

§8 Intuitions about Reference Revisited

We now have in place a full view of how I take the ontology of mathematics to operate. The backbone of mathematical practice, as we have seen, is the algorithmic system. Here is where we find the primary tool of the mathematician, namely, proof. However, we have found that ontological commitments go beyond particular systems. Systems may be, as I have put it, connected, and the objects referred

99 Notice that it doesn't help to claim that mathematical objects are necessary objects or unchanging objects; and that this is why mathematicians are so loath to change the laws that apply to them. For this is to confuse an epistemic fact with a metaphysical one. One is as capable of getting things wrong about objects that are necessary or unchanging as one is about any other sort of object. Or so one would think.

to in the practice of mathematics are referred to by co-referring terms occurring in such collections of connected systems. The core of connectedness is stipulation; that is, all that is required of a connection between two systems, strictly speaking, is that (some of) the terms appearing in one system be assigned co-referrers in the vocabulary of the other system and that, somehow, the potential differences in truth values assigned to sentences taken to appear in the language(s) of both systems be resolved. In practice, such connections will only be made under certain circumstances, and we have taken a stab at describing what character such circumstances must have.

In this last section I want to do three things. First, I want to say a few words about the standard model and its status. The point of doing this is to illustrate the view developed here in a special case: showing what it implies about our knowledge of, and ability to refer to, the standard model.

Second, I want to review the referential puzzles raised in Part I and show how the approach we have developed here solves these puzzles. This, I take it, is to give evidence for the position. I focus on the referential puzzles, and not the epistemic ones, because it should already be clear that the epistemic puzzles can be handled.[100]

Third, I want to compare briefly the view developed here with traditional formalism and with positions that take mathematical objects to be notational. My view, although close in some ways to these, has definite advantages that I want to note explicitly.

We, therefore, start with the standard model of first-order PA. Two questions naturally arise here. The first is the old question of how mathematicians can refer to the standard model because the axioms of PA hold in so many other (nonstandard) models. On our view, this question is easy to answer. PA is just stipulated to so refer. This stipulation explains how the terms of PA co-refer with terms in other systems, systems located in a collection of connected systems that contain some pretty powerful members – namely, second-order PA, ZFC, and so on. Relative to ZFC, for example, wherein the models are defined, the standard model can be distinguished from other models; but due

---

100 Here is a recapitulation for the reader who wants one: Recall that a pair of epistemic puzzles was raised in Part I, Sections 7 and 8. The first was the traditional puzzle about how we could know about mathematical objects, given what they seemed to be. This puzzle was defused by noting that nothing in mathematical practice required mathematical posits to carry the sort of epistemic burden that the traditional puzzle presupposed. Indeed, the crucial point is that mathematical posits have no epistemic burdens at all. They are, as I have called them, ultrathin; and this insight solved our new puzzle too, namely, the question of what epistemic role, if any, mathematical posits have (answer: none).

Instead we find that their role is entirely grammatical, and all that nudges them beyond the simple role of being referents for placeholders in a system is that we desire them, for the sake of pragmatic convenience, to be transsystemic.

to our initial stipulation, we simply take the terms of PA to "really be" about *that* model. For philosophers who worry that this move has all the virtues of theft over honest toil, the rejoinder is simply that when it comes to ultrathin nonlocal posits, there *is nothing* that corresponds (ontologically speaking) to honest toil.

There is a second question not often remarked on, however, that is a little more puzzling. All the systems so connected – PA, one or another axiomization of second-order PA, ZFC, and so on – are incomplete. Consider interpreting the language $\mathscr{L}$ of first-order PA in the standard model in the standard way (e.g., the successor symbol refers to the successor relation, etc.), and call $Th(\mathbb{N})$ the set of sentences of $\mathscr{L}$ that are true in this model. This theory is taken to be a complete one, that is, every sentence of $\mathscr{L}$ can be assigned a truth value based on whether or not it appears in $Th(\mathbb{N})$.[101] How does this happen?

A simple answer that springs readily to mind is that one can *prove* it. One can define the standard model in ZFC, define the language $\mathscr{L}$ of PA, define the notion of satisfaction needed, and prove that $Th(\mathbb{N})$ is complete.

But this answer still seems to leave us with a puzzle. How is it that we have characterized $Th(\mathbb{N})$ within an incomplete theory? Let me sharpen the problem this way: Imagine that we add additional axioms to ZFC. Presumably, because ZFC is incomplete, we can do so in two different and incompatible ways that will enable us to derive *different* (and incompatible) sentences as belonging to $Th(\mathbb{N})$. In what sense, then, have we gotten a grip on $Th(\mathbb{N})$?

Here is one way one could imagine doing so: Consider a hierarchy of systems ascending above ZFC, where each member of the hierarchy, (1) is a metasystem of the system below it and (2) possesses axioms strictly stronger than those of the system below it, so that one can derive new sentences in the language of ZFC (that we take to be true of ZFC because of, say, considerations involving truth predicates) that are not derivable in the system below it. If some such hierarchy of systems, when restricted to the language of PA embedded in ZFC, could be shown to converge upon $Th(\mathbb{N})$, then we would have a fair sense how the set of systems so connected forces the result shown. Unfortunately, there is no such result, and it is hard to see how such a hierarchy that does the job could be established without going quite beyond the bounds of *systems*.[102]

---

101 $Th(\mathbb{N})$ cannot belong to a system because it is not recursively enumerable.

102 Notice that the set of sentences in each of these systems is recursively enumerable. One can define sets of sentences that converge on $Th(\mathbb{N})$, for example, the set of sentences V_n, in the language of PA that contain at most n occurrences of any logical operator, and hold in the standard model. But such sets do us no good.

What conclusion, then, should we draw here? Here's a possibility: For all the apparent slightness of the idiom of truth – its apparent closeness in Tarski's hands to a redundancy theory – in the right company it has great power. In particular, when blended with certain stipulative connections between systems, it forces us to affirm the existence of a standard model, and a set of sentences holding in such, although there is nothing in our actual proof-theoretic practices that compels these affirmations.

I think this way of putting the matter places too much weight on the notion of truth and, consequently, is misleading. Rather, what should be said is that bivalence,[103] plus the definition of the standard model, plus the definition of satisfaction, plus certain stipulations to the effect that the referents of certain terms in PA are the same as the referents of certain terms in ZFC, enable us to prove the completeness of $Th(\mathbb{N})$.

Furthermore, one should add that, in a sense, there is no surprise here. Systems, and the languages they are couched in, are themselves mathematical objects, and this becomes explicit, for example, when one does metamathematics. But, in classical mathematics (which is bivalent), existence proofs that fail to supply details about the objects shown to exist are commonplace. We, therefore, should expect nothing different when it comes to systems.

The philosopher, however, must beware of assuming that these results impose epistemic and metaphysical obligations – that one must explain how the mathematician has gotten access to a set of sentences, namely, $Th(\mathbb{N})$ – without any proof-theoretic means of doing so. Rather, the philosopher must simply recognize that this is something that is the result of adopting the particular set of connected systems we have adopted, a set of systems that involves, among other things, bivalence. The philosophical burden is discharged by recognizing how this came about: The standard model, and the first-order sentences that hold in it, insofar as they go beyond the proof-theoretic methods used to describe them, are mere mirages of co-referentiality and bivalence, aspects of classical mathematical talk that are introduced for reasons of convenience.

If mathematicians, in their desire to further characterize $Th(\mathbb{N})$ (and other sets of sentences like it), connect current systems with stronger ones (say, by adding powerful new cardinal axioms to ZFC that enable them to derive new results in the language of PA), that is their privilege. But no explanation is needed for how talk of $\mathbb{N}$ and $Th(\mathbb{N})$ in the earlier system fixed what the terms in the new system

103 See Quine 1981b.

pick out beyond pointing out that the identification is a simple stipulation.

But isn't there a danger here? After all, aren't (some) mathematicians searching for further facts about the objects of $\mathbb{N}$, for example, for more members of $Th(\mathbb{N})$? Yes, but neither $\mathbb{N}$ nor $Th(\mathbb{N})$ plays a role in this search. For what are such mathematicians *actually* looking for? Merely a stronger system that has the virtues described in Section 6 – and nothing more is needed to explain mathematical practice here. Thus *these desires* explain what mathematicians hunt for, *not* the verbal ghost $Th(\mathbb{N})$. If such a stronger system is found and it is conservative (for current results are appealing), one will simply *stipulate* a connection between our current system and the new one. We describe the situation as one in which we have discovered new tools for learning more about $\mathbb{N}$, and the new results in the language of PA that are now derivable will be taken to belong to $Th(\mathbb{N})$.

Let us turn to the second topic I want to cover in this section, namely, the set of semantic intuitions discussed in Part I. Here is a recapitulation of the intuitions discussed in Sections 4–6.

(1) Recall (Part I, Section 4) that there are strong intuitions that mathematical terms refer successfully right from their introduction into mathematical discourse and, consequently, do not seem to have their references fixed by axiom systems.

(2) Next, recall (Part I, Section 5) that there were a few puzzling intuitions regarding mathematical singular terms. First, the distinction between primary A- and 'A'-mishaps, which is so robust in the case of empirical proper names and singular terms, wavers peculiarly in the mathematical case. Either we find that primary A-mishaps seem intuitively impossible or, in cases where people do have an intuition that a primary A-mishap is present, the distinction between that mishap and one or another 'A'-mishap cannot be made out. Furthermore, intuitions about the presence of primary A-mishaps can often be dislodged with a little work, which is not the case with the corresponding intuitions in the case of empirical singular terms. Finally, one often finds systematic use/mention errors at work with mathematical singular terms.

(3) Lastly (Part I, Section 6), kind terms in mathematics have interesting intuitions associated with them too. Although the distinction between A- and 'A'-mishaps survives here, we find peculiar limitations on the possible extensions of mathematical-kind terms. Their extent seems stipulated in certain respects (when contrasted with empirical-kind terms); for example, there are certain sorts of discoveries about the extent of mathematical-kind terms we cannot make, although no such limitations seem present with empirical-kind

terms. Also, certain mathematical-kind terms seem to shift in their extensions over time in ways that are in tension with the intuitions described in (1). For example, although the functions that 'function' refers to during the time of d'Alembert are included among those that the term now refers to, it seems that there has been a shift in the extension of the term that suggests such an inclusion is problematic: Analytic formulas, even if they are abstract objects, do not seem to be subsets of $\mathbb{R} \times \mathbb{R}$.

The intuitions mentioned in (1) fit well with the picture that has been developed here. Ultrathin posits don't carry much referential weight in any case, and so axiom systems are not required for picking out what such terms refer to (which is not to say, of course, that they don't have other values). Furthermore, because connections between systems are stipulated, there are no requirements on determining whether terms in later systems refer to what the terms in earlier systems refer to. This makes it easy for terms initially introduced in mathematical discourse to refer right from the start.

Ultrathinness also explains both the peculiarities involving A- and 'A'-mishaps and the use/mention errors that are so perennial when it comes to mathematical singular terms. Although mathematical terms are nonlocal, all this amounts to is that terms in other systems can be used to refer to the same objects. There are no other ways of gaining access to the objects referred to by mathematical terms except by the manipulation of sentences containing these terms. This explains why intuitions about the presence of A-mishaps are so wavering, and why it is easy to confuse a particular mathematical object with a term denoting that object: Reference is largely a matter of having singular terms that are treated as denoting terms, and singular terms used this way are carried across systems, and their referential powers with them. The effect that the existence of canonical notation has on such intuitions is also explained on this picture, for the terms in a canonical notation are just terms occurring in systems that are particularly central to mathematical practice – at least as far as those objects (such as numbers) canonically referred to are concerned.

Finally, the disanalogies between empirical-kind terms and mathematical-kind terms become quite explicable on this picture. Something external helps fix what empirical-kind terms are supposed to refer to. But nothing external is available to help fix the extensions of mathematical-kind terms.[104] Consequently, what is available in the system in use at the time becomes much more relevant to what

104 It is at least logically possible that what could be taken to fix the scope of a term like 'number' would be the scope mapped out for that term by systems to be adopted in the future. But, apart from my objections to this move mentioned in Part I, Section 4, intuitions in this area rule out that possibility. Mathematicians

mathematical-kind terms refer to.[105] Notice, also, that given how systems are connected, the tensions one finds between the intuitions that mathematical terms refer right from their introduction into mathematical discourse and intuitions that metamathematical views, to some extent, fix the extensions of mathematical-kind terms are just what one would expect. For in stipulating a connection between two systems, one can often introduce a shift in ontology while *simultaneously* identifying terms in the two systems as co-referential. No problem arises from this because, in practice, the shift in ontology is ignored or not noticed since it is designed to leave the results of the earlier system that mathematicians are interested in untouched. One will be puzzled only if one is a philosopher or historian of mathematics and doesn't understand what is happening here.[106]

The points just made may remind the reader of the problem posed in Benacerraf 1965, and so I should briefly discuss it again.[107] His problem is this: There is a collection of possible embedding of the natural numbers into set theory. All of these are quite compatible with mathematical practices regarding numbers, although they are not compatible with one another. The worry is an ontological one: Which set-theoretic objects are actually the numbers? The picture I

seem unwilling to stake the references of such terms on future practices, and historians of mathematics seem to take this perspective for granted. Thus the extension of a term like 'number' in the hands of the Greeks does not contain complex numbers.

Let me add that I should stress 'help fix'. In general, I do not believe that reference to external objects is straightforwardly fixed by the external objects so referred to, in the case of empirical terms. The picture is more complicated than that, but I cannot get into it now.

105 'Available in the system in use at the time' should be understood fairly broadly. If a new sort of construction enables mathematicians to generate singular terms for particular objects of a kind *K* that could not be picked out before, although the objects fit neatly into the original characterizations of *K*, the extension of *K* will not be understood to have shifted. On the other hand, if the new objects referred to don't neatly fit the previous characterization, then the extension of *K will* be taken to have shifted. This, I think, explains the difference in attitude toward the impact of integral and infinite series notation on our capacity to pick out "numbers," as opposed to the impact of complex number notation on our capacity to pick out "numbers."

106 As we've noticed, what historians of mathematics do is simply note parenthetically that the extensions of particular kind terms are quite different in the hands of different mathematicians, or during certain periods in history, without allowing that fact to impact at all on their practice of taking singular terms used at that time to co-refer with singular terms in current mathematics. This historical practice rarely leads to trouble precisely because connections between systems are chosen by mathematicians in such a way as to maximize the accumulation of mathematical results.

Imagine, however, the mess that would result if historians of (empirical) science tried to do something like this.

107 See Part I, Section 6. I call this Benacerraf's *problem*, which is distinct from what I have called Benacerraf's *puzzle* in Part I, Section 8.

have developed here is quite compatible with the fact that there are many possible ways of embedding the natural numbers in set theory, but one may feel it doesn't tell us what attitude we should take toward this situation. Benacerraf's problem, for example, has been taken to motivate a structuralist view of, at least, numbers. Number terminology refers not to particular objects but to structures that may manifest themselves now here, now there. Is this a view we are going to adopt too?

I am not going to concern myself with every possible motivation available for wanting a structualist perspective. We have seen that such a view is natural for certain branches of mathematics such as group theory. The question is, Does the fact that there is usually more than one way of embedding one system in another *require* such a view?

I think not. Once we recognize how slender mathematical posits are, we should realize that it doesn't much matter if the choices we make in how we identify vocabulary involve fundamental arbitrariness. In practice, those who embed number theory in set theory choose one or another embedding. The fact that there are many available is no more of a problem than the fact that in supplying a coordinate system for a plane, one may place (0, 0) anywhere. One is worried only if one thinks that one must *fix* what number terminology refers to, that one must find what it is out there that, for example, '0' refers to. But this is a fatuous problem, and easily recognized as such by the fact that mathematicians don't spend any time worrying about it.

Here is a way of thinking about what is going on here that doesn't require anything as ontologically radical as structuralism. Early on we saw that mathematics was arbitrary in that one could postulate any system one wanted and study it. Now we find a related sort of arbitrariness, but one that is much less significant mathematically. Just as there is alternative mathematics because of the variety of systems available, there is alternative mathematics because of the variety of connections between systems available.[108] Some of these constitute genuine and interesting branches of mathematics: If one embeds PA in set theories significantly different from ZFC, one may get significant differences in the new results in the language of PA that one can show. On the other hand, many of the variety of connections possible offer only trivial differences: The cases Benacerraf considers fall into this category. Here, too, we are faced with alternatives, but because they make no difference to what can be shown, the distinction is not

108 The spread in "mathematical space" of alternative mathematics, therefore, is constituted not by a variety of different systems but by a variety of different *collections* of connected systems.

noticed. Philosophers, however, should not try to resolve such differences by finding something in common among them to which number terminology, say, can be taken to refer; there is no point in doing this anymore than there would be a point in trying to treat all systems as somehow being about the same things.

One last observation about how this approach handles reference. Recall that worries about inscrutability are inappropriate with local posits. Given that the connections between systems are stipulated, nonlocal posits pose no greater puzzle for reference than local posits do. The reference of 'A' to A is stipulated by the creation of a system, and this stipulation is simply carried along when we connect that system to another. There are no puzzles about inscrutability, for there is nothing out there to scrute.

I conclude this section, as promised, with a brief comparison between my view and other certain views that seem close to it. First we consider traditional formalism. This doctrine is often associated with the following picture: Mathematics (or that part of it that is formalistically reconstrued, anyway) is a collection of rule-governed sentences that, strictly speaking, are meaningless. They may purport to have terms in them that refer, but in fact, this is false. Such collections of sentences are introduced into mathematics because they are valuable in one of several ways; for example, they make the deriving of results that are not formalistically understood easier. Now there is a lot to my approach that makes it look like a version of formalism, and indeed I am quite sympathetic with the motives behind formalism. But ultimately, it is somewhat oversimplified to construe this view as formalist, and here are the reasons.

First, notice that I do not think of mathematical statements as uninterpreted. For one thing, they have truth values. Now it might be thought that for me to invoke truth as a reason for regarding statements as interpreted is fairly disingenuous, for I have stressed that the truth predicate is just a device for asserting groups of sentences without explicitly listing them. Such a device is hardly evidence for one's taking the sentences to be meaningful, if asserting is something a formalist can do (surely Hilbert "asserted" some sentences he regarded as meaningless, and not others).

Let me concede this point and observe that in the classical case, I am perfectly willing to use Tarski's approach to truth. My fears about using it in general, recall, turned on its not being in the spirit of certain systems – its importing, for technical reasons, resources that were intrinsically alien to the system it was being applied to. But where there is no problem with that, I welcome the approach. Tarski's apparatus allows us not only to speak of truth but also of reference (one can say things like, " '1' refers to 1," in the metalanguage), and on the view of many philosophers, the use of it is enough to show that one is not taking the sentences it can be applied to as meaningless.[109]

Here's a response to these comments that someone who still wants to insist that the position developed in this book is formalist is likely to make:

I'm going to give you the notion of "understanding," and agree that talk of formalistically construed sentences as being "meaningless" on the part of formalists like Hilbert was a bit much. If someone plays a game like chess, it is hardly the case that the pieces are meaningless. On some views, all that is needed for something to be "meaningful" is that the rules governing its use be clear and (perhaps) that we have certain psychological responses when we use them. Mathematical terms are meaningful because there are rules governing their use, and it is also worth noting that we have certain psychological responses to such terms (certain images spring to mind). Nevertheless, one is still a *formalist*, in some sense of the word, if the terms one uses fail to refer to anything. Furthermore, the mere invocation of Tarski is insufficient to prove that one takes one's terms to refer. True, some philosophers, such as Davidson and Quine, take Tarskian reference to be all the reference one has. But it is certainly reasonable to think that talk of reference on the Tarskian approach is entirely dependent on the view one takes of one's terms in the object language. If there is no reason to think that the terms of one's object language refer, then there is no reason to think this of their co-referrers in the metalanguage, despite the fact that one talks in the metalanguage of the object language terms as "referring." The term 'linguistic realist' (used in the opening pages of this book) is a misnomer, for on its own linguistic realism is not realism at all. More is needed.

If someone takes this view of the Tarskian apparatus, then perhaps I am, to him, a formalist.[110] But one more aspect of my position must be considered. This is that I allow terms to co-refer across systems, even when the rules governing these terms in the respective systems *are not the same*. That is, we are allowed to say that we got something wrong about an object, and that the rules governing the term that referred to it in an earlier system are wrong. This sort of thing hardly seems to be in the spirit of traditional formalism.

Now, it is true that I explain what is going on here not in terms of our somehow recognizing that the *object* referred to didn't have the properties we thought it had but, rather, in terms of certain later systems having virtues preferable to those of earlier systems. And, one

---

109 In particular, Benacerraf (1973) takes it this way. But he is hardly alone in doing so.

110 Although I must confess that the position, once one concedes the meaningfulness of the sentences of mathematics, seems to have been reduced to a simple version of nominalism.

might say, for the mathematical object to have dropped out like this on one's picture *is* for one to treat mathematical terms as irreferential.

At this point, it seems, there is no way for me to avoid being called a formalist, if that is what someone wants to do. I don't mind, provided it is realized that an interesting constraint has been placed on our notion of reference: Ultrathin posits (and thin posits, for that matter) cannot be referred to. Or, to put it another way, there are no ultrathin (or thin) posits. Terms that refer to such things are fictional.

I have no idea how one would argue for (or against, for that matter) such a position; but to call me a formalist, or to deny I am a formalist, would be to take a stand on that issue.[111]

One last point: One cannot help noticing, I suppose, that the systematic use/mention errors I describe in Part I are ones that, although explainable on the approach I have taken, are most directly handled by taking mathematical terms to be *about* notation. Why haven't I taken this approach? Well, first notice that to do it honestly would require quite a bit of reconstruction. Physically realized notation is probably not available,[112] and one must, therefore, either (i) understand such mathematical notation platonistically (which leaves us in regard to the epistemic puzzles of Part I exactly where we started), (ii) introduce a possible-worlds semantics in order to make the referents for one's notation available in nearby possible worlds if they can't be found here, or (iii) consider falsifying a bit of (classical) mathematics. This means that more than a direct explanation of use/mention errors is needed to motivate such a position. And if I am right, no such motivation is available: Taking our terms to refer to ultrathin posits solves all the philosophical problems, it leaves mathematical practice entirely intact and entirely reasonable, and it leaves the semantics for mathematese analogous to that of the rest of our language. Why bother, then, reconstruing mathematical language as really about itself?

111 See the end of Part III for a related, brief (and perhaps equally unsatisfying) discussion of whether the position I have developed here is a nominalist or a platonist one.
112 On current physical theories, anyway.

PART III

The Geography of the A Priori

But I have found no substantial reason for concluding that there are any quite black threads in it, or any white ones.

W. V. O. Quine

§1 Introduction[1]

Part II was primarily concerned with the ontology of mathematics, although epistemological issues were never far off stage. However, in this final part of the book, we will be almost exclusively concerned with what are, broadly speaking, epistemological issues. We take up three interrelated topics: a priori truth, the normativity of mathematics, and the success of applied mathematics. All three of these topics have had a substantial presence in philosophy. For example, the success of applied mathematics played a significant role in motivating Kant 1965. Similarly, one or another notion of a priori truth has played a substantial role in the epistemology and metaphysics of many philosophers, even to this day. Finally, philosophers of mathematics have seen the normativity of mathematical law as a deep fact that rules out what might otherwise be appealing positions; for example, consider the use of it made by Frege and Husserl in their attacks on psychologism.[2]

By contrast, I have fairly deflationary views on a priori truth, the normativity of mathematics, and the success of applied mathematics; in particular, I believe that not much of philosophical interest follows from a close examination of any of these.

Let us first consider a priori truth. Doctrines of a priori truth were motivated by the felt perception of a difference between the epistemic properties of mathematical truths and those of nonmathematical truths. Such doctrines fell on hard times with Carnap, who attempted to force a priori truth into the Procrustean bed of meaning relations, where it nearly died (at the hands of Quine).[3]

Philosophers have subsequently come to distinguish apriority from analyticity; and the question of whether or not Quine succeeded in demolishing the analytic/synthetic distinction has generated a formidable literature. However, Quinean views have pretty much carried the day as far as a priori truth is concerned, especially when it comes to mathematics; and exploring the epistemic continuities between mathematics and the empirical sciences has become quite popular subsequent to Quine's first voicing his opposition to the possibility of a sharp distinction here.[4]

Worries *have* been raised, even by philosophers who largely accept Quine's picture. For example, we find Putnam (1983f) arguing that there is at least one particular truth that must be a priori, and we find

---

1 Some of the material contained in Part III appeared in earlier forms in Azzouni 1990 and Azzouni 1992. As with Part II, I have not necessarily respected either the identity, contours, or content of those earlier essays.
2 See Frege 1894, 1967 and Husserl 1970.
3 Quine 1951a.
4 A sample: Lakatos 1976a, Kitcher 1984, Putnam 1975a, Tymoczko 1979, Crowe 1988.

Parsons (1979–80) arguing that Quine's way of explaining away a priori intuitions (via the centrality of mathematical truths in the conceptual scheme) won't do for mathematical truths that are particularly obvious. But the demise of the Carnapian attempt to found a priori truth on meaning relations has left philosophers who take this sort of line in a curious position: They can certainly offer examples of a priori truths that are intuitively convincing to most of us, but they have no general story to tell about how such truths, in Kant's phrase, are possible.[5]

My aim is to steer a course between the reviving of the old-fashioned view of a priori truths as constituting a substantial part of the bedrock in one or another epistemic foundationalism, on the one hand, and the assimilation of mathematics to the empirical sciences, on the other. Although, that is, I think I can make out a distinction between empirical science and mathematics, and shall try to mark it out by something like a notion of a priori truth, the distinction in my hands will be far less dramatic (epistemologically speaking) than the traditional one (s) seems to have been. In particular, a priori truths, in my sense, will be neither incorrigible, obvious, nor independent (in any sense) of empirical experience. They will, however, turn out to be independent of empirical *science* – and this will go far toward explaining why knowledge gathering in the mathematical sciences is so different from knowledge gathering in the empirical sciences. It also explains why I am willing to adopt the traditional nomenclature for my nontraditional concept.

The basis of my notion of a priori truth is the algorithm which I discuss in Sections 2 and 3. The notion so defined proves to be valuable because of its fairly central role in mathematical practice, as I illustrate. In Sections 4 and 5, my aims are a bit more destructive: I am out to supply an explanation for commonplace intuitions that certain simple mathematical truths (such as '1 + 1 = 2') are obviously true. The traditional notion of a priori truth was designed, in part, to explain these intuitions (or at least was widely seen to be supported by such intuitions), and because my new notion cannot do this job, I must explain them in some other way. In Section 4, I do so by focusing on the pervasive use of certain patterns of inference codified in simple mathematical truths, and noting that these are necessary for us to, for example, execute algorithms. It is the perception of this pragmatic need that I take to be behind the intuition in question. In Section 5, I turn to Putnam's attempts to revive a pre-Quinean notion

5 Parsons considers reviving a version of the positivist story. Putnam hopes to ground a priori truth in a notion of rationality. The first move does not look promising to me. The second seems explanatorily idle.

of a priori truth–that is, truths that we cannot be wrong about – and I show that there is no reason to think there are *any* truths of this sort. Whatever is special about the knowledge afforded by mathematics and logic, I claim, it is not that incorrigible truths may be found in those domains.

The upshot is that the central use of (and need for) algorithms in mathematics suffices to explain what is (and seems to be) special about mathematical truths.

As the reader will see, despite my deflationary views on a priori truth, I have found myself with a lot to say, both positive and negative, about one or another version of the notion. This is not quite the case with the two other topics I mentioned in the beginning of this section. For example, some people may find it amazing that applied mathematics works so well, and may feel, as Kant did, that a rather deep explanation is needed here. I am not one of these. Applied mathematics is successful only when it cavorts with one or another empirical science, and my suspicion is that its success has a rather humdrum explanation quite similar to the kind of evolutionary explanation we offer for the success of various species of animal. I give that sort of explanation here.

Finally, there is the matter of the normativity of mathematical law. The tendency of philosophers is to see this as marking something special (and deep) about mathematical laws, in particular, their irreducibility to empirical laws. I see normativity, by contrast, as having its source in the ontology of mathematics and therefore as being due not to a deep property of mathematical law but, rather, to a shallow property of the objects of mathematics. In particular, normativity will turn out to have its source in obvious methodological facts (already discussed in this book) about what the best way to do mathematics is.

There is one other aspect to normativity in mathematics that I will have something to say about, and this is the question of what the nature of mathematical errors is. This matter will turn out to be a little more subtle than one might have expected.

Recall that we have made a distinction between cases where someone makes a mistake about what sentences belong in a system Γ and cases where a person is actually using the rules of a different system Γ^* correctly. We need a systematic explanation of how such a distinction is made. The reason this problem is connected to the *normativity* of mathematics is that, if the question of what system a person is using is purely descriptive, then there seems to be no sense in which a person can make a mistake: Whatever that person does *must be* in accord with the rules of the system he or she is using. The normativity of mathematics, in this sense, lies in the fact that there is a potential gap between what a person's mathematical practices are and what the correct practices are – and the capacity to make mistakes in a practice lives off of this gap.

We will find that normativity in this sense has roots in two distinct phenomena. First, there is the general practice, already remarked on, of connecting systems. Just as we have found that what connections exist between systems is a matter of stipulation,[6] so, too, whether early practitioners are subsequently seen as having made mistakes about the nature of the objects they were studying will be a matter of stipulation.[7] However, this is not the whole story about the fine art of blundering. Even aside from how systems are connected to each other, mistakes are possible; and even when mathematicians are (relatively) clear about what the rules governing their practices are, they make mistakes. In this case, as we will see, the attribution of mistakes to practitioners will be a little more robust, turning on systematic distinctions found in the general practices of the mathematical community.

§2 Algorithms Again

We start with a few preliminary observations. It is natural to think of justification in the mathematical context, proving theorems or perusing proofs, as practices that can be, and are, carried out by individual mathematicians on their own. This gives the following picture of knowledge gathering in the mathematical profession as a whole: Mathematician A proves a certain theorem, which is then published or passed around informally (either by word of mouth or in manuscript form). Other mathematicians read the proof, or learn about the result and produce a proof on their own. In this way, the theorem makes its way through the profession until everybody who is interested in the subject has worked out the proof one way or another for him- or herself. Justification of mathematical results *for the profession,* therefore, is taken to *supervene on* acts of justification carried out by members of the profession on their own.[8]

We routinely attribute knowledge to groups, however, without that knowledge supervening on justificatory practices of all the members of that group. I go on to discuss two examples of this sort of thing.

First, it can happen that only some of the individuals in a group have actually justified a particular belief directly for themselves,

---

6 Part II, Section 6.

7 To adopt a pregnant term of Putnam's, the normativity of mathematics will be somewhat a matter of *projection*. See Putnam 1987.

8 Of course a mathematician may fail to see how one step follows from another and ask for help. But this doesn't change the picture in any essential way, because the explanation given must be one he or she *understands*.

and the justification of the belief for everyone else rests on these individuals. All of us have many beliefs that we hold solely because we have heard them from sources that, for one reason or another, we trust. I should add that, in many cases, one doesn't accept testimony unless one has justified for oneself that it has authority, where, by *authority*, I mean that the information contained in the testimony can be traced back to evidence methods that are sound ones. But this is hardly the normal case, as Descartes realized long ago. As students, as children, even as people reading newspapers, we are rarely in a position to determine whether the individuals informing us have authority.[9]

Second, it can happen that the process of justifying a particular item of knowledge is not something that is carried out by any *one* particular individual. Rather, there is a division of justificatory labor at work; no one individual has directly justified the belief.

Here is an illustration of this second possibility. The belief that there are atoms is not something that any one individual, no matter how trained in physics, can directly justify for him- or herself. Rather, the evidence for this truth is a complex combination of mathematical derivation and empirical data that no one person has ever perused completely.

Testimony is present in this second sort of case, too, but in a way that is different from our first case. In the first sort of case, recall, A may believe a fact on the basis of B's testimony, and B believe it because she has justified her belief directly. However, in the second sort of case, A may believe a fact on the basis of B's testimony about certain aspects of that fact plus what A has done himself, while B can be in exactly the same position with respect to A (although in regard to *different* aspects of the fact in question). Here, too, I will say that the justification for a particular belief for an individual relies (in part) on *testimony* when it relies (in part) on what the individual has heard or read or otherwise gleaned from others; and as with the first case, I will describe the testimony as having authority when it can be traced back to evidence methods that are sound ones.

One of the things I will show in this section is that group practices of justification in which testimony plays an ineliminable role occur in mathematical practice to a far larger degree than has been hitherto realized, and this must be taken into consideration when describing the epistemology of mathematics. Despite this, however, it will turn out that there is still a sharp distinction between the evidence-gathering practices in the mathematical sciences and those available

9 If the "media culture" we live in is a healthy one, the testimony we receive does have authority (most of the time anyway). But this, alas, is largely a matter of epistemic luck. I cannot get further into this very interesting topic now.

in the empirical sciences; in particular, a nice working notion of a priori knowledge will still have a central place in the epistemology of mathematics.

I start by constructing a notion of a priori knowledge that is relevant to the mathematician as an isolated knower; and only after indicating the scope and limits of this notion in mathematical practice do I turn to those epistemic practices among mathematicians as a professional group that do not supervene on individual knowers.

To begin with, it is often pointed out that it is not entirely appropriate to speak of *truths* being or not being a priori, because even if a truth could be known a priori, it doesn't follow that it *has* to be known a priori. After all, A could simply tell it to B, and A could have enough authority that B now knows that truth as a result. Therefore we shall apply the term 'a priori' not to statements but rather to *epistemic warrants,* where that phrase is meant to describe a legitimate *process* of justifying that a statement is true.

With this terminology in place, let us give the following:

Definition 1: An individual A has an *a priori warrant* for the claim that a sentence α belongs to a system Γ (an α/Γ-claim) if A has surveyed a derivation of α in Γ.

A caveat: Obviously, if A has surveyed a derivation, someone or something must have created it. I allow the creator to be either a mathematician or a computer.

A second caveat: I am setting aside the details involved in requiring in some way that A not only have surveyed the proof but also *understood* the implications of what he or she has surveyed. Some clause to this effect is needed in order to prevent certain bizarre counterexamples to the definition. I am simply going to take this requirement as a necessary condition on surveying a proof, for example, on someone's having fulfilled the conditions of Definition 1, without in any way trying to spell out how it goes. Similarly, I am waiving niceties about how recently the person must have surveyed the proof, or how much of it he or she must remember in order to still possess the warrant. These are interesting issues, but I have nothing particularly illuminating to say about them now.

Yet a third caveat: Definition 1 presumes that what is invariably involved in the justification of an α/Γ-claim is a token of the actual proof of α in Γ. But this is not quite right. What is often given, rather, is a proof sketch. A proof sketch, if it is sound and valid, can be turned into a genuine proof (although sometimes this is quite difficult). Proof sketches are best understood as extrasystemic, enthymemic arguments that proofs exist. But our confidence that a particular proof sketch really shows that the proof it sketches exists relies on the same sort of evidence that comes into play when evaluating a real proof, and so we can modify Definition 1 implicitly to allow a priori warrants even when what is possessed is not an actual derivation but only a sketch of one.[10]

Having defined 'a priori warrant' this way, I will eventually go on to give reasons why the process of coming to know a mathematical truth via a derivation is epistemically special, and special in a way that makes it reasonable to call the process an "a priori" one. Before doing so, however, I need to make clear that there are at least two ways in which such a process is *not* epistemically special. First, I do not see the process of creating derivations as one that goes on independently of experience, however that phrase is to be understood. The reason is this: In general, to recognize that a particular sentence belongs to a system, an algorithm must be applied to it. Now, although certain algorithms can be carried out mentally, and although certain very gifted individuals are quite capable of carrying out even quite complex algorithms mentally, most of us require pens and paper, chalk and blackboards, fingertips and calculators. Furthermore, it is clear that, in general, the computations needed to determine whether a particular sentence belongs to a particular system are quite open-ended, so that, invariably, auxiliary helping devices will be needed.

Epistemically speaking, this means that the recognition that a particular sentence belongs to a particular system is one that owes its epistemic warrant, in part, to whatever justifies our confidence in the empirical devices we use to help ourselves carry out the algorithm. I simply see no way of gainsaying this, and so, if we take a priori warrants as discussed, there is no sense in which we can regard them as independent of experience.[11]

Second, a priori warrants, in the sense given in the definition, do not license *incorrigible* beliefs in the α/Γ-claims established. In general it is always possible to make mistakes in the execution of an algorithm, even when the steps involved are quite short. Mistakes, over time, are invariably ironed out, but still they are always possible. So another property traditionally associated with the a priori is absent here.

10 It will become obvious eventually that substituting proof sketches for proofs changes none of the epistemic qualities I attribute to a priori warrants in the sense of Definition 1. See, in particular, note 26.

11 It doesn't follow, of course, that one should add to the axioms of every pb-system an 'Axiom of Symbolic Stability', as Schröder did (see Frege 1953, p. viii^e). To do this would be to *stipulate* the good behavior of the world around us.

There need be nothing manifestly obvious about whether a particular sentence belongs to a particular system.[12]

Because a priori warrants are not connected to processes that are epistemically independent of (external) experience and do not license incorrigible beliefs, what makes them epistemically special?

The matter is delicate, but there *is* a distinction to be made here, one between an epistemic dependence on empirical *experience* and an epistemic dependence on empirical *science*. This is a distinction that will not be welcome to certain philosophers, so I have to argue for it.[13] Here goes.

Consider the process of making marks on paper. It is true that such a process need not have been a reliable one. For it could have been that, now and again, paper affects the marks on it so that they change in ways that make attempting to carry out an algorithm with pencil and paper a mistake (or something that leads to mistakes often enough to warrant avoiding doing it). How have we ensured that this doesn't happen?

Well, we have used *pre-scientific* methods of measuring the reliability of the products (such as paper and pencils) that we work with. By this I mean that we have methods that were available to us long before the flourishing of empirical science. I also mean (and mean to show) that such methods are actually epistemically prior to the results in empirical science.

A beautiful illustration of how such methods work is the measurement of time. We have no direct access to the flow of time itself.[14] All we have are various processes by which we can measure the passage of time, processes such as the movement of the sun, the slow burning of various substances such as wax, rope, and so on, the flow of substances from one place to another, such as sand in an hourglass, and the subjective sensation of time at a crawl or time in a rush. What is

---

12 However, doesn't it seem that certain very simple mathematical and logical truths *are* manifestly obvious – that one *couldn't* be wrong about them? We will come to this eventually. See Section 4.

13 The distinction is generally overlooked by philosophers who deny the possibility of a pure observational language. For example, consider this quotation from Lakatos: "[C]alling the reports of our human eye 'observational' only indicates that we 'rely' on some vague physiological theory of human vision" (1976b, p. 107). Although such reports, as he claims, are not purely observational, this is not because some vague physiological theory of human vision is involved. The epistemological facts have been reversed here. Any theory we develop must support the veridicality of what we see (at least most of the time) on pain of undercutting its own evidential basis – or so I will argue shortly. Furthermore, surely many people (and cultures) have never even considered the possibility of a theory of human vision, let alone relied on a vague one. Other philosophers with the same tendency easily come to mind: Quine (e.g., his 1960), Feyerabend, and Churchland 1979. There are also philosophers of mathematics who blur this distinction, for example, Resnik (1989).

14 This is, perhaps, a strange way of putting the matter, but bear with me.

required of these various processes is *regularity*. But regularity is not something that can be recognized any more directly than the passage of time itself. Instead, what must be done is to zigzag our way toward regularity by comparing the rates of various processes against the rates of other processes. As it turns out, temporal irregularity in such processes is not coordinated,[15] and consequently we can recognize the more regular processes by way of contrast with the less regular. This is a paradigmatic example of how, pre-scientifically, we evaluate the reliability of the materials and processes around us.

The world could have been so unfortunate a place as to yield epistemic slippage everywhere. It could have been that the sort of gross materials and processes that we work with in our daily lives, that we *have* to work with for biological reasons (for we have access only to so much via our senses), and that we have worked with since before the dawn of civilization could have exhibited the sort of instability that one finds occurring at the atomic level.

Luckily this is not so, for otherwise we would not know anything.[16] Now here is the point. Because the world offers us enough stability to know things, it offers us enough stability for us to be able to *check* the reliability of certain methods of recording data against other methods.

Here is an example. I have developed confidence in floppy disks because what is recorded there has matched printed copy when I check it later. This confidence has nothing to do either with scientific doctrine (for I know virtually nothing about how floppies work) or with the authority of experts (for I do not know any experts in this area). It is simply because I have checked the reliability of floppy disks in a pre-scientific manner – that is, against other methods I have for recording data – that I am convinced that they are reliable.

That these methods (of checking our epistemic claims about such objects and processes as we have direct access to) are independent of, and prior to, justification in the sciences can be recognized by observing that the idea that scientific developments could somehow show that such methods are systematically misleading is virtually incoherent. For *were* such methods misleading, we would have no way of testing scientific claims either; all such claims are tested, *and can only be tested*, by transforming them into processes that involve indicators of the sort that must be checked in a pre-scientific way. For example, if positrons act in a certain way, that is evidence for or against certain theories. But such evidence is of no use to us until it is translated into

15 Well, if it is – for example, if God periodically "stops the clock" – we don't notice anyway.
16 I am assuming for the sake of illustration that life would be possible under such circumstances, although perhaps this is false.

a form that we have access to: Certain lights must flash. And, in turn, to be sure that we have seen what we think we have seen (a red light flashing several times, and not a green light flashing once), we rely on pre-science.[17]

A caveat: Someone might complain that I have stated that we are stuck with pre-scientific methods without showing that they are epistemically *justified*. But what could possibly be wanted here? One is tempted to simply say: There are our methods. In any case, I cannot imagine what other ones would look like, for, as I have already suggested, if the world were so unpleasant that these methods did not work, we could not know anything.[18]

Another caveat: By no means am I suggesting that such methods provide anything like the incorrigibility philosophers hoped to glean from sense-data languages. What I have called pre-scientific methods are fallible ones for sure; but their fallibility is in terms of the pre-scientific tests we bring to bear on them. For example, I might realize, by checking data storage in floppies against other methods of storing data, that floppies are actually *more* reliable than paper and pen, even though I was first convinced of the value of floppies by comparing them to paper and pen. But it barely makes sense to suggest that, by bringing *scientific theory* to bear on paper and pen, we could show that such devices are actually quite unreliable – without, that is, this unreliability *already* manifesting itself in some way that I could recognize *without* the application of scientific theory.

A third caveat: Holists might claim that scientific development could subsequently come to justificationally underwrite our prescientific insights into the reliability of the gross processes and materials we work with. I am entirely in agreement; indeed, to *some* extent this goes on. But my point is that from an epistemological point of view this underwriting is idle inasmuch as such pre-scientific insights cannot be overthrown by scientific development.

The general picture, therefore, is this: Empirical justification is two-tiered.[19] The bottom tier is composed of our pre-scientific methods for testing reliability and gathering evidence. This tier is largely epistemically autonomous, and justification there operates more or less according to the standard coherentist picture. The second tier is composed of empirical scientific knowledge. It is also coherentist in

17 One might think that talk of relying on *pre-science* is a bit grand. Don't we just look and see? Not necessarily. For we may use collateral information about how things look to us under certain circumstances. In a room with a blue light, we may recognize that certain colored lights do not look as they are supposed to. See the caveats forthcoming.

18 If my opponent is still not satisfied, it appears that the request for justification comes down to the requirement that one refute Cartesian skepticism. *Maybe* that is required, but we can't get into it now.

19 I owe this felicitous term to Stephen White.

epistemic structure, but foundationalist in that the evaluation of all such knowledge is supported by methods drawn from the first tier.[20]

I take these arguments to show that there is an epistemic distinction that can be made between cases where our surety of something depends on pre-scientific methods of justification and cases where our surety depends on the truth of certain scientific doctrines.

With such a distinction in hand, it is clear that a priori warrants in the sense of Definition 1 are grounded not in scientific doctrine but in pre-scientific doctrine: the evidence of our senses and a certain confidence in the stability of the world (at least as it bears on chalk and blackboards, pen and paper, and so on). Results with a priori warrants enjoy a certain insularity from change in scientific doctrine. Relativity theory has come, and someday it may go, but its presence or absence will not impact on the results of algorithmic processes. This does much, I think, to explain the intuition that mathematical results are necessary results; for this means, among other things, that such results are more robust than the results of empirical science. We cannot go so far as to treat the process of establishing mathematical results as one that is independent even of external experience, but we must regretfully regard *that* doctrine as simply a matter of overkill.

We turn now to the issue of what role a priori warrants in the sense of Definition 1 play in the mathematical profession as a whole. But before addressing this, a preliminary second definition is called for, namely:

Definition 2: An a priori warrant for an α/Γ-claim is *possessed by a community of mathematicians* if a derivation of α in Γ is surveyed by members of that community on a regular basis.

A caveat: I am leaving aside messy details about how many members of the mathematical community should survey the derivation in question and exactly what "a regular basis" comes to. Generally speaking, if a derivation is widely available in textbooks, or if most mathematicians studying an area the derivation is relevant to have easy access to it, I shall take the conditions of Definition 2 to be satisfied.

The important issue is this: How much mathematical knowledge possessed by the mathematical community at any one time is actually covered by Definition 2? It might seem that quite a lot is. After all, mathematicians spend their lives producing proofs of theorems (or

20 There is a *lot* more to say about two-tiered coherentism, and a lot more to offer by way of argument for it. In particular there are interesting issues here about the role of sense perception in pre-scientific practices and its vulnerability to biotechnical modification. But this must wait for another time because, except for the application I make of this view to a priori warrants, the discussion would be a digression.

attempting to do so, anyway) and reading proofs by other mathematicians. It is standard for papers and texts of mathematics to be organized in what is largely a theorem/proof format.

This impression is easily dispelled, however, if one asks what *systems* these theorems are theorems of. Consider Quine 1940. Presumably every theorem in that book is a theorem of *ML*. But it hardly seems to be the case that one's knowledge of this fits Definition 1, or that the knowledge that these theorems belong to *ML* on the part of the (sub) community of logicians familiar with *ML* fits Definition 2. This is simply because the derivations of results toward the middle of the book or the end of the book (e.g., †331 on page 188) are not surveyable.

Let me expand this point a little by means of this example. The reader, upon looking up Quine's proof of †331, may think that I have greatly exaggerated the difficulty here. For one need only trace the proof back to its assumptions, trace those assumptions back to their assumptions, and so on. The proof (in this particular case), although a fairly long one, can *hardly* be described as unsurveyable, and the same seems to be true for every other theorem Quine proves in the book.

I concede the objection but don't think it affects my point. For one can hardly read the book by tracing *every* proof back to the axioms in the way described. And, in practice, if one wants to read a book like this from cover to cover, that is not what one does. Rather, one relies on the memory of one's earlier reading to stop the regress pretty much where Quine stops it, that is, in the citation of assumptions shown previously.[21] But to suggest that this practice in some way involves an actual survey of derivations of theorems from an initial postulate basis seems crazy. For such a survey requires not the mere tracing back of assumptions for a particular proof in the book: It has to be a tracing back for *every* proof in the book, and that really is a monstrous task. For consider: Either one, in the process of surveying, is constructing a giant branching tree of derivations that one can draw assumptions from, when required to, for a new theorem, or one is constructing a giant linear derivation where one remains aware, in some sense, of the location of salient results along the derivation. Neither process is psychologically plausible, and consequently neither can be how someone reading Quine 1940 is actually justifying how the theorems in question belong to *ML*.

One might think that Quine's book, like any book in mathematical logic (around that time, at least), presents a relatively isolated phenomenon. For more stress was laid in mathematical logic, then, on

---

21 Actually, one doesn't usually rely on one's memory of the earlier proofs; one relies more simply on the recollection that one was satisfied with what one read when one read the earlier proofs.

derivations in a formal sense than one found (and finds) in mathematics proper. This, although true, supports my contention. Consider as an example the graduate text Hewitt and Stromberg 1965. One does not find there the dense indexing of assumptions characteristic of Quine 1940, but this hardly implies that the authors have constructed their proofs from scratch. Instead, the assumptions and definitions at work are far more complex than those in Quine 1940, and, consequently, on pain of unreadability, the authors simply take most of them for granted. Theorem proving here has become much more of a matter of *indicating* to the already mature student what should be salient to him or her; the extensive background, which would be explicit in any genuine derivation, is not articulated. Ironically, therefore, the lighter indexing of assumptions in a text like Hewitt and Stromberg 1965 does not indicate an easier task of surveying derivations from an initial postulate basis; rather, it indicates that the task is irredeemably intractable.[22]

It is claimed (and for pretty good reasons) that most mathematics can be captured in ZFC. Presumably this is to be shown, if it can be shown, by *deriving* the results of this mathematics *in* ZFC. But the vast majority of these results are not available in surveyable derivations from the initial axioms of ZFC, for these *really would* run eventually to thousands of pages.

When one turns to research papers, one finds something similar. A proof that α is a theorem of Γ is rarely the exhibition of a proof or proof sketch of α from a postulate basis of Γ. Rather, it is a proof or proof sketch of α from a set of implicit and explicit statements that are already known to hold in Γ. In other words, strictly speaking, a research paper usually affords a priori knowledge in the sense of Definitions 1 and 2, not of α's belonging to Γ, but of its belonging to Γ^*, where the latter system has as its postulate basis the set of statements and inference rules implicitly and explicitly used as premises and inference rules in the proof or proof sketch of α exhibited in the paper.

Any systematic textbook has the same structure. To return to our initial example, what is contained in Quine 1940 is not a gigantic derivation of a list of theorems from *ML* but, rather, a patchwork of der-

---

22 The impression one gets from such books is that what the student is supposed to be learning is how to manipulate mathematical concepts in an almost instinctive way. When definitions of mathematical terms are offered, it is not that one is supposed to keep in mind the primitive terms such definitions are built out of; rather, one manipulates the terms directly, using definitions *and anything else one has learned about the objects in question*, until one has a good sense of how they operate.

This is not to claim that something is going on here that cannot be captured in terms of derivations from a postulate basis for a mathematical system such as ZFC; it is to say that such derivations are constituted at a social level in the profession. The individual mathematician, for the most part, operates in a (derivationally speaking) more inchoate situation.

ivations of theorems from different systems, all of which are recognized (although not necessarily in an a priori sense) to be subsystems of *ML*.

Recall that we spoke earlier of *testimony* in cases where someone's knowledge of something relies on what the knower has heard or read from others. Such cases as we are considering now involve testimony in this sense; that is, we understand a mathematician to be relying on testimony when he or she takes a system, Γ^*, to be a subsystem of another one, Γ, because the author clearly intends this (e.g., the book is clearly about Γ in the way that Quine 1940 is clearly about *ML*), or because the author shows this in an earlier part of the book that the mathematician is not concerned with, or because the author cites an assumption in a derivation as a well-known result in the area, and so on.

Now there is a clear sense in which the testimony involved here can be eliminated *in principle*. Here is an illustration of how this is possible. Consider a particular system Γ and consider a complex proof of α in Γ by a mathematician A. Suppose that in proving α, A uses as premises $\beta_1, \ldots, \beta_n$, all of which she has gotten from textbooks or from papers, but none of which she has actually examined the proofs for. Now if all is well in the mathematical community, we can find individuals (possibly none the same) who have generated or checked the proofs for $\beta_1, \ldots, \beta_n$, to which the testimony that A relied on can be traced. These proofs in turn may rely on testimony as to additional premises $\gamma_1, \ldots, \gamma_m$, which in turn must be traced back to their proofs. Eventually, this process must terminate so that a proof could be carried out (a proof that has been generated or checked by no one individual in the group, as it turns out) in which testimony plays no justificatory role.[23] Only if this is possible will we describe the testimony supporting an α/Γ-claim as *having mathematical authority*. With this piece of terminology in place, we have a last

Definition 3: We say that a community of mathematicians is *mathematically justified* in its collective belief that an α/Γ-claim is true if either the community possesses an a priori warrant for the claim or the testimony supporting the claim has mathematical authority.

Now I was vague about exactly what the upper limits on a surveyable proof are, and I was also vague on whether a proof perused over the course of several days (or months, for that matter) constitutes a survey of that proof – for I have offered no arguments, one way or the other, for whether one can take up a proof later, rely on one's

---

23 One can imagine, as I've suggested, that the division of labor is so intense that for reasons of time and energy no one individual could actually either check or generate the proof in question. So the 'could' is highly idealized.

memory, and yet still be said to have surveyed the proof in question when one is done. It should be clear that I am willing to understand 'surveying' quite broadly, because I conceded that I did not require of my notion of 'a priori warrant' that it have certain traditional virtues, such as independence from empirical experience or incorrigibility.

One objection, therefore, to my suggesting that rather little mathematical knowledge on the part of the mathematical community fits Definition 2 might take the form of claiming that someone who has read Quine 1940 from cover to cover *has* surveyed (if we are sufficiently broad-minded about what constitutes "surveying") every theorem of *ML* shown by Quine. This might raise the worry among some philosophers that the considerations that moved us against using Definition 2 alone as a definition of *mathematical justification* are not that significant. For if it is possible, in principle, to eliminate the use of testimony in the justification of any α/Γ-claim, why shouldn't we regard the uses of testimony in this arena simply as epistemic shortcuts, concessions to practical constraints on time and energy? Why not regard an α/Γ-claim as mathematically justified *only* when a derivation of α in Γ has been surveyed?

A philosopher will want to make this move only because he or she thinks there is a significant difference in the epistemic quality of justifications based on surveying derivations as opposed to justifications based on testimony with mathematical authority. Let me make two observations that will undercut this impression.

First, notice that the crucial epistemic facts having to do with prescience are still in place when we shift from mathematical knowledge held on the basis of surveyable derivations to that supported (in part) by testimony. This is simply because the knowledge that I presuppose in taking account of the authority of testimony in this area is no more justified by empirical science than my presuppositions about the stability of chalk on blackboards. That is, when I am convinced of a certain result on the basis of reading an abstract by a certain mathematician A, this is not because (a) I am in command of sufficient scientific knowledge of how A's brain works to trust his capacity to generate correct derivations, and (b) I am in command of sufficient knowledge of the science of sociology to trust the mechanisms by which that knowledge has been transmitted to me. Therefore, the gain in surety by actually surveying the derivation in question is not enough to make the profession treat the cases as epistemically different.[24]

24 Indeed, it is clear that the profession does not treat results derivationally remote from the premises of a postulate basis for a system as having a different epistemic status from those results derivationally close to those premises, unless the distance is so great that only a computer can bridge it. See the discussion of "computer-generated proofs" forthcoming.

In all fairness to the opposition, however, it is worth mentioning that surveyability certainly seems to be a desirable virtue, and is seen as such in the profession. For mathematicians spend almost as much time refining proofs for results already known among them as they do generating new results. It is seen as quite appealing, in particular, to show how to derive a number of laboriously achieved results in a neat and quick way.[25]

This is not a dull matter of derivation bookkeeping. It shows that surveyability is an epistemic virtue much sought after in mathematical practice (it is a primary, although not sole, component of "elegance"). I think the epistemic desirability of surveyability needs no explanation. I want only to stress that the epistemic gain it affords lies totally within the domain of pre-science: It is not the case that an increase in surveyability here affords an escape from the reliance on empirical science, as no such escape is needed, nor that it affords an escape from empirical experience, as no such escape is possible.[26]

The second observation I want to make is that mathematicians, professionally, rarely go through a mathematical text in the step-by-step manner that the text is written in. Rather, they pick and choose what they wish to look at, both when it comes to particular proofs in a text and when it comes to the steps in a proof itself; and their choices in this matter are dictated by their interests as well as their background knowledge and understanding.[27] Consequently, it is rather unusual, in practice, for a mathematician actually to have surveyed derivations for all the α/Γ-claims he or she is taken to know.

Let me bring this section to a close by first contrasting briefly the view developed here with the rather influential view developed in

25 An example philosophers are apt to be familiar with is how the Löwenheim – Skolem theorem and the compactness theorem for first-order logic fall quickly out of the proof for the completeness theorem. But the phenomenon is absolutely rampant in mathematics.

One should not overrate, however, how much the practice of providing new proofs for old results indicates the significance of surveyability for the profession, for, usually, new methods are involved in the new proofs, and often it is these new methods that are significant because they can be applied elsewhere in order to derive *new* results.

26 This should make it perfectly clear why there is often no epistemic loss, and in fact sometimes an epistemic gain, in surveying a proof sketch rather than the actual proof. Length costs in surety: A short but perspicuous derivation sketch may make it *more* obvious that a derivation exists than the actual derivation itself, especially if the details of the latter can cause us to lose touch with what is going on.

27 I was once told by a professor of mathematics that his opinion of someone's mathematical ability would drop sharply if he learned that he or she had read a mathematical text from cover to cover. The crudeness with which the sentiment was expressed was perhaps unusual, but not, I believe, the sentiment itself.

For a while, the field of mathematical logic was an exception, for it operated with a different paradigm of "rigor," but I am under the impression that this is no longer the case.

Descartes 1931a, and then by considering "computer-generated proofs" in the light of the distinctions made here.

Descartes' view requires that for one to be sure of a deduction it must be "scrutinized by a movement of thought which is continuous and nowhere interrupted. . . ."[28] Surveyability, that is, becomes the only method by which one truly justifies that one knows that A follows from B. It may be inappropriate to saddle Descartes with the view that what is required is not the surveying of a proof sketch but a survey of the actual derivation itself in my technical sense; but when he complains in his discussion of Rule VII that people have a tendency to skip steps, it is hard to see why anything less than the derivation itself will do.

On my view, surveyability is a virtue to be much sought after; but on the view I've attributed to Descartes, *most* of what mathematicians do simply does not get them knowledge of what they take themselves to know. Furthermore, the advice that Descartes offers for how one is to make sure that one indeed has produced a good proof is advice that many mathematicians would scorn (running the steps of the proof continuously over in one's mind until one has, as it were, an intuitive perception of the whole thing is dull beyond belief). On producing a proof she is not sure of, a mathematician more likely will show it to a colleague she trusts for confirmation.[29] On Descartes' view, and natural elaborations of it, such a procedure is epistemically crazy.

I draw the conclusion, not that I have unearthed a growing epistemic crisis in mathematics, but that the picture of justification in mathematics is not anything like what Descartes described.

Let us turn now to "computer-generated proofs." My distinction between an epistemic dependence on empirical experience and an epistemic dependence on empirical science bears directly on the status of such proofs, and here's how. The practice of mathematics predates serious empirical science by several centuries and, in particular, predates our capacity to bring our scientific knowledge to bear on mathematical theorem proving by a couple thousand years or so; until recently, one could plausibly argue that *all* epistemic warrants for mathematical truths were mathematically justified in the sense of Definition 3. But now our recognition that certain sentences belong to certain systems turns not on the actual exhibition of a proof, the checking of an already generated proof, or on a belief held on the

28 Rule VII, Descartes 1931a, p. 19.
29 Sometimes, mathematicians are more sure of the result than they are of the proof. Proofs, that is, are often patched up, even at the point of publication. This is only possible because a mathematician's confidence in the result has not been achieved in the way Descartes requires.

basis of testimony with mathematical authority, but on whatever epistemic surety is available that certain computers have successfully generated the proof. Computers, that is, can supply empirical existence claims for mathematical proofs, existence claims that, because of the limitations of human capacity, cannot even be checked *by the mathematical community as a whole* by examining the proof the computer has purportedly generated.[30]

Now I want to stress that mathematically unjustifiable computer-generated proofs are really not proofs at all.[31] They are empirical demonstrations that certain proofs exist, and, consequently, although this need not stop them from using such results, mathematicians do have a more slender epistemic grip on such things than they do on proofs that are actually possessed by the mathematical community.

To illustrate this, consider the following thought experiment. Imagine that a certain mathematician proves that a counterexample to the four-color theorem exists. What will happen? I predict the following. First, his proof will be scrutinized very carefully for errors. But suppose after quite a bit of scrutiny, the proof is recognized as sound (suppose it is actually not that difficult a proof to check).

At this point, I imagine that mathematicians will attempt to scrutinize the computer-generated proof, at least in its general outlines, and also consider the details of certain cases. Some, I imagine, might even make the study of this computer-generated proof their life's work. Suppose, though, nothing turned up (although the search continued for several generations).

There seem to be two choices. The first is that one could draw the (tentative) conclusion that mathematics (that part of it, anyway, that these proofs occur in) is inconsistent. The second choice would be to assume that the computers (for several have been involved, checking each other's work) have made a mistake. I claim that most mathematicians would draw the conclusion that the computers have made a mistake, and that they would henceforth take computer-generated proofs very much less seriously.[32]

I admit that this is only speculation. But notice that the intuitive evidence my thought experiment yields supports, not the centrality of mathematics to our conceptual scheme in the Quinean sense, but rather our greater confidence in our own capacity to generate math-

---

30 See Tymoczko 1979 and the articles cited therein. Also see Detlefsen and Luker 1980.

31 I will continue to use the term 'computer-generated *proof*' in the case where the purported proof is not mathematically justifiable, but understand it somewhat the way one understands terms 'purported criminal' or 'forged Picasso'.

32 Computer scientists would think otherwise, I imagine, if they could not determine how the computers malfunctioned. But *mathematicians* would not agree.

ematical results. It points, in fact, to the distinction between empirical science and empirical experience that I have argued for.

The conclusion is this. Until the advent of computer-generated mathematical work, I claim that theorem proving was, epistemically speaking, independent of empirical science. This can and may change: It may come to pass that many mathematical results are not mathematically justified in the sense of Definition 3, for there is no reason in principle why computers couldn't come to play an integrated and ineliminable role in theorem proving. However, two factors are at work that would still sharply separate epistemic practices in mathematics from epistemic practices in the empirical sciences:

(1) Due to the nature of how mathematical results are established, it will always be possible in principle to separate those mathematical results that have a priori warrants or are held on the basis of testimony with mathematical authority from those that don't and aren't. Consequently, it will always be possible to separate sharply mathematical results that we are more sure of (because they are mathematically justified in the sense of Definition 3) from those we are less sure of. Confirmation holism has no place in the establishment of mathematical truth as we understand that process to operate.[33]

(2) The second factor is one that I am reminding the reader of, for I have already spoken of it at length: The epistemic incursion of empirical science into mathematics only occurs at the level of evidence gathering. The process of recognizing that a particular sentence α belongs to a system Γ, as we've seen, may be one that is dependent on computer science. However, in no way has the epistemic purity of mathematics established in Part II been compromised. It is still the case that what must justify the legitimacy of an epistemic process that establishes an empirical truth is some reliability connection between the process of confirming such a truth and the objects referred to by such a truth. No such requirement holds for the process of confirming mathematical truths.

§3 Some Observations on Metamathematics

We have cashed out a notion of the a priori in terms of Turing machine computability, a mathematical theory that has been applied to our very own algorithmic practices. In choosing to apply this mathematical theory to these practices, we have applied a mathematical subject to an empirical subject area (the theorem-proving practices of a certain kind of animal). This means that the distinctions made here

---

33 The classic presentation of confirmation holism occurs in Quine 1951a. The view is that sentences are not confirmed or disconfirmed individually but only meet the test of experience as a group.

are distinctions in an empirical science; in particular, the distinction between what is a priori and what is not a priori in this sense of 'a priori' is an empirical distinction: It turns on the mathematical theory of Turing machine computability being *true of* our own algorithmic practices.[34] But this makes the characterization of the a priori dependent on empirical science! Of course it has not been shown that what is a priori is actually not a priori (i.e., that what we take to be independent of the empirical sciences is actually not independent of the empirical sciences), because it doesn't follow that a property of a concept is automatically a property of what it is a concept of. Nevertheless, I suspect this kind of argument is disturbing enough to motivate giving the entire issue a closer look.

A branch of (naturalized) epistemology[35] is the study of algorithms, because they are a tool for knowledge gathering. Naturalized epistemology (as a whole) is an empirical subject because it is the study of the study habits of a particular kind of animal. For example, if what I have argued up until now is right, a priori warrants do not depend on results from the empirical sciences: This is a result in naturalized epistemology.

But this result doesn't prevent us from creating an applied theory of algorithms that arises from the empirical study of our practice of mathematics itself. Metamathematics is taken to be applicable to the practice of mathematics, but its being so applicable depends on the empirical sciences (in particular, on naturalized epistemology).

A few observations are required to eliminate the impression that metamathematics poses a puzzle for my views on pure and applied mathematics. Here's the apparent puzzle. I have characterized the branches of pure mathematics as subjects ontologically independent of empirical science, and I have spoken of positing in this context, suggesting that such positing is not much weightier epistemically than the quite similar positing that goes on in games. But metamathematics seems to operate differently: It seems to have as a subject

34 Every mathematical theory is true; we have seen that already (see Part II, Section 4). But not every mathematical theory is *true of* every particular empirical domain, where we take a mathematical theory A to be "true of" an empirical domain B if we take A to apply to B. Whether a mathematical theory is true of – can be applied to – a particular empirical domain is an empirical matter.

35 The term 'naturalized epistemology' carries with it a whiff of controversy. For some, the term denotes a branch of science, perhaps psychology, which is supposed to replace or supplant the traditional philosophical subject; whereas for others it is an oxymoron, the name of a subject that in some sense is supposed to be continuous with the traditional subject but approaches it in such a way that its philosophical content is hopelessly compromised. See Quine 1969c, 1969d, Goldman 1986, Dretske 1981, Putnam 1983g, and Stroud 1981, among others.

I cannot discuss this complex issue here. I want only to say that my use of the term 'naturalized epistemology', although friendly to Quinean usage, does not shortchange the traditional subject in the way that some philosophers fear. But, unfortunately, this is not a claim I can justify now.

matter our mathematical practices *themselves*. If true, this carries a lot of implications. In particular, the nature of proof turns out to be an empirical matter, something we might regard as implicitly recognized when "Church's thesis" is so called. For in calling Church's thesis a thesis, one suggests that it could be *wrong* – that is, that it is possible we could discover effective procedures, procedures that could be incorporated into our methods of proof (or perhaps are already informally there), that are not Turing computable. How, then, is metamathematics to be a branch of mathematics?

I think the situation here is almost analogous to that of geometry. We distinguish between geometry practiced as a pure mathematical subject – where there is no question of whether the results correspond in any way to physical space – and geometry as an applied subject. The latter gives rise to the (empirical) question of which (pure) geometry, if any, is best applied to physical space. Sometimes this question is expressed in the form, "which geometry is true?" but the position developed in Part II, Sections 3 and 4, rules out that way of expressing the question as quite misleading. Another way of putting the question, which is not, to my mind, so misleading, would be to ask, "what geometric structure does physical space have?"

In the same way, there are alternatives to computation theory based on Turing computability. On the one hand, we could consider theories in which Turing computability is supplemented with oracles. On the other hand, there are weaker models of computability such as finite-state machines, top-down automata, and so on.[36]

As it turns out, no one of these alternative notions best captures human computation. Turing computability is widely regarded as constituting the upper bound on our capacity for computation – and this, I suspect, is what is behind calling Church's thesis a thesis. For we don't have a *proof* that Turing computability constitutes the upper bound on human computational power, and it is hard to see how we *could* have a proof of this fact. But, of course, this is not all there is to the question of human computation. For Turing computability ignores various constraints that human computation labors under: Limitations of time and materials is one example, and our general incapacity to identify extensionally equivalent functions is another; and these limitations are absolutely crucial to the question of which computations are feasible *for us*.

36 But these are not seen as *alternatives* in the way that non-Euclidean geometries were seen as alternatives to Euclidean geometry. Fair enough. As branches of mathematics, non-Euclidean geometries are no more alternatives to Euclidean geometry than computation theory based on finite-state machines is an alternative to computation theory based on Turing machines. On the other hand, as far as the applicability of various alternative geometries to physical space is concerned, they are *as much* alternatives as the various computational possibilities are in regard to the question of which model of computation best captures human computational capacities.

Thus the theory of computation, based as it is on Church's thesis, is a mathematical theory in which Church's thesis operates as a simple premise. In *applying* this theory to human computation, Church's thesis functions differently: It is a substantive assumption about the upper bound on *human* computation.

There is one subtlety here that does not arise in the case of alternative geometries, and it must not go unnoticed. In speaking of unapplied metamathematics, I am thinking of a subject that is unapplied *empirically*. That is, I am thinking of a subject in which there is no concern with whether anyone can actually do mathematical proofs of the sort permitted. One could (and does) consider metamathematics where oracles are permitted (e.g., an oracle for recognizing the sentences that hold in the standard model of PA).

But there is another sense in which the mathematical subject of metamathematics *is* applied, and that is to the mathematics that it is metamathematics for. This subject arises by identifying certain systems in the way described in Part II, Sections 6 and 7, but must be distinguished from the empirical subject that I am also (misleadingly) calling metamathematics, where the question of what methods of proof in mathematics are available *for us* is crucial.[37]

The puzzling feel of the arguments here is at least partly due to the fact that we use a branch of mathematics in our study of the practice of all branches of mathematics (including itself). Indeed, such self-referential studies are necessary if the practice of epistemology is to go on within the conceptual scheme itself. For, apart from despairing of the practice of epistemology altogether, the only other option is to step outside our conceptual scheme. And where would *that* leave us?[38]

§4 Incorrigible Co-empiricalness

Quine recognized early on that his attempt to nudge the mathematical sciences in the direction of empiricism (epistemologically speaking) faced the requirement of explaining (away) strong contrasting intuitions. One such intuition is that mathematical truths are incor-

---

37 Standard textbooks in this area tend, in my opinion, to conflate the two subjects, and the entire issue is usually buried by calling 'effective computability' an *intuitive* notion. But this leads to the sort of objection made by Mendelson 1990, that Church's thesis is just like any number of other identifications in mathematics between "intuitive" notions and other more formal notions, and should not be dignified with the term 'thesis', as if in some way it has not been established. I think Mendelson is right to make this complaint, but feel that what logicians such as Kleene and Church were getting at by calling Church's thesis a thesis can be made clear by distinguishing the mathematical subject of metamathematics from the application of it to our actual mathematical practices.

38 Presumably in some *other* conceptual scheme. But epistemologists should not find the regress looming here appealing.

rigible, and Quine's explanation of this intuition invokes the *centrality* of mathematics and logic to our conceptual scheme.[39] That is, he argues that the usage of mathematical and logical truths is so ubiquitous, that revising them would impact so widely on everything we believe, that only enormous benefits could motivate us to do so. Centrality, thus understood, is a pragmatic constraint on conceptual schemes. Because Quine looks to theory as *the* source of centrality in this sense, let us call it 'theoretical centrality'.

Now some philosophers dislike using theoretical centrality to explain away intuitions of incorrigibility; this is especially true of those who feel that the intuition subsists not merely on resistance to the revision of such truths but on the recognition that the absence of such truths would result in an *incoherent* conceptual scheme.[40] But the explanation is certainly well aimed at the intuition in question.

Not so for another intuition closely related to the one just mentioned, that of the *obviousness* of (some) mathematical truths, an intuition that Parsons notes cannot be explained away by the presence of centrality.[41]

What exactly does the impression of the obviousness of statements such as '1 + 1 = 2' come to? I detect three factors. The first is that such statements are true. But in the light of Part II, this does not amount to much, as truth has turned out to be, mathematically speaking, a very cheap commodity. Second, and this is rather more important, there is an intuition that such truths apply to empirical experience (i.e., to the gross materials and processes we meet with in our daily lives) and to the empirical sciences. It is not merely that such statements are true, but that they are *true* of the empirical world. I am going to describe the intuition this way: '1 + 1 = 2'; and other truths like this, are *co-empirical*.[42]

I have stressed so far *what it is* that is supposed to be so obvious about these statements (after all, it could have been *obvious* that they

39 Quine 1936, p. 102.
40 See, for example, Katz 1979 and Putnam 1983f. I discuss these intuitions in Section 5.
41 Parsons 1979–80, p. 151 is worth quoting on this: "The empiricist view, even in the subtle and complex form it takes in the work of Professor Quine, seems subject to the objection that it leaves unaccounted for precisely the *obviousness* of elementary mathematics (and perhaps also of logic). It seeks to meet the difficulties of early empiricist views of mathematics by assimilating mathematics to the theoretical part of science. But there are great differences . . . [such as] . . . the existence of very general principles that are universally regarded as obvious, where on an empiricist view one would expect them to be bold hypotheses, about which a prudent scientist would maintain reserve, keeping in mind that experience might not bear them out. . . ." Parsons also makes the significant observation that "the obviousness of logic is an unexamined premise of [Quine's theory of translation]."
42 I am tempted to merely call them 'empirical'. But given how I have been using that term in this book, that would be extremely misleading.

are false). But what is important here is not only that such statements are true and co-empirical but that they are obviously so – so obviously so, in fact, that it seems we have *incorrigible knowledge* that such statements are true and apply empirically.[43] I describe the situation by saying that we have the intuition that such sentences are *incorrigibly co-empirical*.

It may seem surprising to suggest that part of the intuition here is that such truths apply *empirically*. *Why* would anyone think that our knowledge of *that* was incorrigible? Wouldn't it be *obvious*, rather, that whether or not a truth applies empirically is something up for grabs (something *empirical*)?

I can only say that it seems clear to me that empirical applicability is definitely part of what is supposed to be so obvious about statements like '1 + 1 = 2'. For were I to come forward and offer the perspective on truth given in Part II, Sections 3 and 4, I am sure that this would not be regarded as an adequate reconstrual of these intuitions. Something more is clearly at work, and what could it be but the impression that, of course, these truths apply empirically.

Parsons obviously intends the intuition to have this implication, for he is speaking within the Quinean framework; and there, for a logical or mathematical statement to be true immediately implies its empirical applicability.

In any case, as I have indicated (Part I, Section 11), it seems to me that there are no purely Quinean resources available for handling intuitions of incorrigible co-empiricalness. Let me illustrate this claim in a fresh way with something from Churchland 1979. Churchland distinguishes between the degree of *systematic importance* and the degree of *semantic importance* that a statement can have. The former is the degree of centrality in Quine's sense, and the latter is our degree of dependence on the statement for smoothness and efficiency in our verbal commerce. The latter notion is a social one: The degree of semantic importance of a statement measures how often we use that statement to communicate the usage of a term contained in it to others.

I regard the introduction of 'semantic importance' as a move entirely friendly to the spirit of Quine's approach.[44] But although the notion certainly does something to explain the sensation of obviousness we feel about statements such as 'Bachelors are unmarried males'

43 It may seem odd that the impression of obviousness here involves *both* the truth *and* the co-empiricalness of the statements in question. At times, to speak of something being true is to speak of it as being co-empirical in the sense I mean it. (See, for example, my observation, in Section 2, about the question of which alternative geometry is "true.") But because we are understanding truth more narrowly than that, we must cash out explicitly the other notions at work in the truth idiom when it is being understood in a more substantial manner. This is what we are doing here.
44 Hopefully Quine would agree.

or 'Ice is frozen water', it is less successful for the (obvious) mathematical truths. This is because, I think, there is a very strong impression that the semantic importance of such truths is due to their obviousness and not vice versa. Furthermore, this impression is supported by the fact that many mathematical truths are quite obvious, but never used in the attempt to communicate mathematical concepts to others. Something different is going on.

Now the failure of one example of how one might try to draw distinctions within the Quinean framework to accommodate intuitions about incorrigible co-empiricalness does not show that it cannot be done. But it is very hard to see what other resources Quine has available to do this; and thus it is no surprise that Parson suggests (1979–80, p. 151) going back to a pre-Quinean view about logic being true by virtue of meaning to explain this impression in the case of certain *logical* truths. I am pessimistic about this move; there are simply too many such truths significantly differing in content from one another (which upon inspection seem obviously true) for this to be reasonably taken as a matter of "meaning."

Where, then, are we to look for the source of this intuition? Because I have suggested several components are involved here, let us take them in turn. First consider incorrigibility. What can be behind such an intuition? I claim that it can only have its source in our methods of confirming evidence. If we have incorrigible knowledge about a statement, we must see, given how we confirm statements of that sort, that it is manifestly obvious that we cannot be mistaken. Something in our method of confirming the statement must *guarantee* its truth, if our knowledge of it to be incorrigible. Consider, as an example, Descartes' perception of the *cogito*.[45] The method of establishing this sort of truth is inspection of one's intuitions. And if the intuition is presented to us clearly and distinctly, so the story goes, we cannot be wrong about it.

Now one complication heretofore overlooked is that the description I just gave seems to link incorrigibility to the methods of confirmation we are *currently* using. But one can imagine us dropping certain methods of confirming truths for others. In this case, even if it is manifestly obvious that certain statements must be true, given a particular method of confirmation, this might not be the case if we switch to another method. What seems required for *incorrigibility*, therefore, is not merely that we can't be wrong about the truth of a statement, given current methods of evaluating its truth value, but that we can't be wrong about the epistemic value of these methods *either*.

Illustrating the point with Descartes' *cogito* again, it seems that what is required for incorrigibility is not only that one see, given the

45 See Descartes 1931b.

method of seeing something clearly and distinctly, that the *cogito* holds, but that one see that this very method is one that we cannot be wrong about; we would never come to think that the method is a bad one.

Let us turn to the question of where one is to find sources for the intuition of co-empiricalness. Here what are relevant are restraints on how much empirical science and pre-science can change; the restraints must be severe enough to explain our intuition that certain mathematical truths are necessarily co-empirical. One place to look, although not the only place, for restrictions like this is again our methods of confirmation. If we find that there is a certain rigidity in such methods – for example, if we find that no matter how our methods of confirming truths change, we will always need to add 1 of something to 1 of something, and get 2 of those somethings – then, it would seem, '1 + 1 = 2', anyway, must be co-empirical.

Now the Quinean lesson that this section opened with is that one can sometimes explain away intuitions without taking them at face value. Although *his* focusing on theoretical centrality fails to explain away intuitions of obviousness regarding certain logical and mathematical truths, it doesn't follow that our only choice is to fall back on a picture that accepts these intuitions at face value. As it turns out, I have no intention of claiming that statements such as '1 + 1 = 2' and 'Not every sentence is false' are incorrigibly co-empirical. In fact I will argue that no statement is such.[46] How, then, will I explain such intuitions?

The considerations of the last few paragraphs suggest we look to our confirmation procedures. I think this is right. Apart from Quine's sort of centrality, what I have called 'theoretical centrality', I will invoke an alternative source of centrality in conceptual schemes, one found by focusing on our methods of confirming truths. I will find a certain class of *sentential patterns with atomic truth conditions* to be *evidentially central* because one or another such pattern is required by our confirmation procedures. The reason that this so-called evidential centrality does not guarantee the co-empiricalness of any statement (and consequently our incorrigible knowledge of this sort of fact with respect to any statement) is that the requirement that a class of sentential patterns with atomic truth conditions must be available for our confirmation procedures does not imply a requirement on the part of these procedures for any *particular* statement. Thus evidential centrality is much weaker than the incorrigible co-empiricalness of statements. Nevertheless, it will be sufficient to explain the intuitions discussed at the beginning of this section.

46 See Section 5.

What I have just written will be made much clearer after I take a little time to explain the piece of jargon I have just introduced: 'sentential pattern with atomic truth conditions'.

Recall, from Part II, Section 2, the notion of a deductive relation. Two sets of sentences, Γ_1 and Γ_2 (with a deductive relation defined on them), are *syntactically equivalent* if a deductive isomorphism from one to the other exists, that is, if there is a one–one onto mapping $\Omega: \Gamma_1 \mapsto \Gamma_2$, where given that $\Delta_i, \Delta_j \subseteq \Gamma_1$, then $\Delta_i \vdash_{\Gamma_1} \Delta_j$ iff $\Omega(\Delta_i) \vdash_{\Gamma_2} \Omega(\Delta_j)$. We will also say that one set of sentences Γ_1 *is deductively contained in* another set of sentences Γ_2 if there is a deductive isomorphism between Γ_1 and a set of sentences $\Delta \subseteq \Gamma_2$. (This condition is easily seen to be equivalent to another: that there be a deductive *homomorphism* $\Omega: \Gamma_1 \mapsto \Gamma_2$; i.e., a mapping from Γ_1 *into* Γ_2 that is compatible with the deductive relations defined on Γ_1 and Γ_2).

Now the existence of a deductive isomorphism between two such sets of sentences does not necessarily indicate that the sentences so identified are the same or express the same things. Well, why not? One reason is that the deductive relations (determined by the systems in play) that a sentence has to other sentences is not the whole story about what a sentence means. A crucial part also is under what circumstances a sentence can be asserted, where these circumstances are determined in a way that is external to the implication relations imposed by the system. That is, part of the meaning of a sentence, in general, is the "truth conditions" of the atomic sentences contained in it, where these truth conditions are defined independently of the implication relations.[47] For example, the truth conditions of 'Grass is green', we will assume, connect the statement 'Grass is green' with grass and its properties in such a way that the statement will be assigned the truth value *true* if grass is green. This can be independent of whether the system such a sentence is couched in is classical, intuitionistic, or whatever.[48]

Several caveats: First, I should point out that *my* use of 'truth conditions' is meant to be plural. To explain why, I will indulge in a bit of informal semantics. We imagine that there are (vaguely specified) possible contexts in which statements can be applied and be assigned truth values. Each such context, and resulting truth value assigned to the statement so applied, I take to be a *truth condition for that statement*. This way of putting matters will allow me to speak of 'changing some of the truth conditions of a statement' and mean simply that some of the truth assignments to the statement in certain contexts are being

47 I am waiving technical complications raised by quantification.
48 Of course, one's intuitionism may cut so deep that one will not always agree with the classical theorist over when circumstances hold in which "Grass is green." We'll come to this momentarily.

shifted.[49] I will speak of the *positive* truth conditions for a statement, and mean by this the contexts where the statement is assigned *true* as a truth value.

Now, in the classical situation, truth conditions admit of a neat completeness. One has both positive and negative truth assignments that are exhaustive and disjoint with respect to the set of contexts. But we may have a nonclassical situation to deal with, so I am only concerning myself with the positive truth assignments. These will suffice for our purposes. Indeed, one can speak of assertibility conditions if one wishes; but note in any case that talk of *truth* conditions should be understood in the spirit of Part II, Sections 3 and 4.

Also, in the classical situation, from truth-value assignments to atomic sentences we may derive truth-value assignments for complex sentences by means of the deductive relation defined by the classical system. The process is simply one of deriving complex sentences or the negations of such, using one or another postulate basis for the standard predicate calculus.[50] We might say that the postulate basis here supplies the *derivation conditions;* they fix the truth conditions of the complex sentences in terms of the atomic sentences. Because of completeness, what I have called the derivation conditions are equivalent to the standard Tarskian truth conditions on connectives and quantifiers.

When we generalize to the nonclassical context, we have derivation conditions as before and can use them to go from positive truth conditions for atomic sentences to positive truth conditions for complex sentences. But here, in general, there is nothing corresponding to Tarski truth conditions. For our purposes, however, derivation conditions will do just fine.

Finally, I understand 'truth conditions' as 'true-of conditions'.[51] This makes no difference for the empirical statements, but it enables me to extend the informal apparatus just described to mathematical and logical truths, as well as to the inference rules. On this view, although all mathematical and logical statements are true, they need not be true *of* every context; and although all inference rules are "truth preserving," they need not be truth preserving *in* every context. So I will understand 'positive truth conditions' to be items that are assigned (via assignments to atomic sentences and the derivation

49 I have nothing to say about how my talk of 'contexts' should be formalized in terms of situations, possible worlds, or whatever. The use I shall make of this apparatus is, semantically speaking, so simple that I suspect what I have to say (when construed classically) will be compatible with most of the alternatives available in the literature.
50 This process of derivation, of course, goes not just from the simple to the complex but vice versa. We are not concerned with the other direction, however.
51 See note 34.

conditions admissible in that context) not only to what are traditionally construed as empirical statements but to logical and mathematical statements and inference rules, too.

We are prone, however – and I have for the most part gone along
with this inclination – to take logical truths and inference rules, as
well as mathematical truths, to apply in every empirical context, if
they apply empirically at all. But now we are going to allow the possibility that such statements and inference rules may have positive
truth conditions in certain contexts but not others. There is nothing
intrinsically incoherent about this suggestion. For example, we can
certainly imagine that in certain contexts we would not allow inferences licensed by the standard predicate calculus but only inferences
licensed by the intuitionistic predicate calculus. In such contexts,
truths such as 'Either John is running or it is not the case that John is
running' will not have positive instances (except under special
circumstances).

Let Δ and Ξ be two sets of sentences with deductive relations defined on them, where truth conditions have been assigned to the
atomic sentences in the sets (hereafter, I call the truth conditions assigned to the atomic sentences in a set of sentences, with some inac-
curacy, 'atomic truth conditions'). We say that Δ and Ξ have the same
sentential pattern with atomic truth conditions if there is a deductive isomorphism $\Omega: \Delta \mapsto \Xi$, which obeys the additional condition: If A is an
atomic sentence in Δ, then τ is a set of positive truth conditions for A
iff τ is a set of positive truth conditions for $\Omega(A)$.

If Δ and Ξ are such that there is a deductive homomorphism $\Omega:$
$\Delta \mapsto \Xi$, which obeys the additional condition: If A is an atomic sentence in Δ, then τ is a set of positive truth conditions for A only if τ is
a subset of ν, the positive truth conditions for $\Omega(A)$; then we say that
the sentential pattern with atomic truth conditions exemplified by Δ is con-
tained in the sentential pattern with atomic truth conditions exemplified by Ξ.

The reader may be wondering why two sets of sentences with the
same sentential pattern with atomic truth conditions do not, in general, express the same statements. What more could be needed? Part
of the answer is this. The two sets of sentences may be embedded in
two larger sets of sentences that do not have the same sentential pattern. In such a case we will not always regard the sentences in the two
sets, paired by virtue of playing the same roles in the sentential pattern, as expressing the same statements, because, although their deductive relations to other sentences, when restricted to sentences in
the sets, are identical, what statements such sentences express may
well be a matter decided in part by deductive relations they have to
sentences in the sentential pattern at large. But a more obvious reason is that I am not requiring that the set of positive truth conditions,

τ, in the definition be the *entire* set of positive truth conditions available for a sentence under a particular interpretation. The reason for this generalization is that we will be concerned with the constraints imposed on statements by virtue of our execution of algorithms, and the positive truth conditions involved *there* will not, in general, be all the positive truth conditions for these statements as *normally* interpreted.

Here are two examples. If we restrict our attention to those subclasses of the sentential calculus and the intuitionistic calculus that contain the connectives '∧' and '∨' (as well as an interpreted set of atomic sentences), the two sets of sentences have the same sentence pattern with atomic truth conditions. But it is debatable whether or not classical sentences of this sort express the same statements as their intuitionistic counterparts.

Second, consider the sentence 'This frog is green'. One set of positive truth conditions is all contexts containing an indicated green frog. A narrower set of positive truth conditions is all the contexts mentioned earlier that have occurred in the last five years.

One last piece of terminology, and then we will finally get on with the topic at hand. I will sometimes want to allude to the place in the pattern that a statement is located in, without actually meaning to describe the statement itself. For example, I may want to describe a particular place in the sentential pattern exemplified by 'John is running ∨ Peter is sleeping', where '∨' is intuitionistic, and 'John is running ∨ Peter is sleeping', where '∨' is classical, without meaning to speak of either the intuitionistic or the classical sentence located in that place in their respective systems. In such a case I will have a token of a sentence that either normally expresses a statement or is used to stand for a sentence follow the generic term 'principle', as in "The principle '1 = 1'...." 'Principle' here is functioning as a term of art, marking out something that we use to reason with but which is not localized to any particular applied system.[52] I will also sometimes speak of 'the truth conditions assigned to an atomic principle', when it is open which particular system of sentences is involved; and I will also speak of 'logical principles' and 'mathematical principles', where what I mean are the principles in a sentential pattern that are, usually, exemplified by logical truths or mathematical truths, respectively. Finally, I will sometimes speak loosely of 'inference pattern' when I want to stress the application of the *principles involved* to license steps in a piece of reasoning.

I should point out that my talk of 'principles' presupposes that one can identify roles that sentences have across sentential patterns, even

---

52 'Principle' differs from my other term of art, 'statement' in that, if two sentences express the same statement, then they are taken to mean the same thing. Two sentences that exemplify the same principle need not mean the same thing.

when these patterns differ. To some extent this is intuitively unproblematic, as the commonplace comparison of classical sentences with their intuitionistic counterparts makes clear.[53] In general, the problematicality of this locution varies directly with how different the sentential patterns involved are. I will use the locution without doing much by way of justifying it, for example, giving necessary and sufficient conditions for when one is faced with the same role in two different sentential patterns, contenting myself only with the observation that in most cases we do pretty well in figuring it out if a comparison can be made at all.

We turn now to an explicit discussion of evidential centrality, largely as it concerns our methods of confirming mathematical truth. In Part II, Section 4, I called the logic and mathematics that are used in the empirical sciences and in our daily lives *standard* mathematics and logic. As we did there, we will again take standard logic to be the first-order predicate calculus and standard mathematics to be whatever mathematics can be couched in ZFC; although, as before, nothing will turn on this choice. Such truths are, in the terminology we have just adopted, co-empirical. I want, however, to distinguish the co-empirical mathematics and logic utilized in science generally from that strict subclass of truths utilized in our evidential procedures for the mathematical sciences.

In order to do this, let us consider the evidential procedures used in mathematics generally. To speak of evidential procedures in this case should not seem odd because we have focused so much already on the algorithms we use for confirmation there. Algorithms, of course, range from shorthand calculating tricks to the subtle application of high-powered theorem proving. The important point for us is that it seems that certain empirical and co-empirical truths must be taken for granted to apply algorithms in the first place. Some of these, as we've seen, are ordinary empirical truths about how the mechanisms we use to record our results operate, for example, that chalk dust on slate and pencil marks on paper are fairly stable. But most important for our purposes is that some of these are mathematical and logical truths.

Consider the example of verifying a tautology via a truth table. In such a verification, it is routine to *count* rows of 'F's and 'T's. Simi-

53 This practice can lead to awkwardness, however. We find Dummett (1977, p. 35) writing: "Intuitionistic arithmetic is not, in practice, very different from classical arithmetic: that is, there are few theorems to be found in textbooks of classical number theory which cannot be proved, sometimes after minor reformulation, in intuitionistic number theory," after explicitly arguing for a difference in meaning between *all* the logical particles of classical logic and those of intuitionistic logic. Clearly, he must be using 'theorem' in a way that abstracts away from the specific (and different) meanings of the theorems as they appear in intuitionistic and classical number theory, although he makes no attempt to explain his usage.

larly, in order to carry out certain proofs, one may have to count cases, steps in a proof, quantifiers, or even parentheses! Notice what is absolutely crucial here. One *is not* using simple theorems about numbers as *premises* in the proof of a certain result. To assume that is to commit precisely the sort of error that used to be made by those who objected to the logicist definition of 'There are five Gs', that one needed to be able to count quantifiers in order to recognize that the definition was suitable! Nevertheless, the epistemic point is a sound one, *at least as far as humans are concerned.* Numerical facts play an unavoidable epistemic role when we apply such algorithms; without applying such facts to our algorithmic practices, we could not show the results we need to show.

Notice, too, how being able to apply these sorts of algorithms the way we do requires simple logical inferences to hold too. This can be seen by the fact that numerical inferences are never pure; logical inferences are always involved as well. Another illustration is this: When one lists the rows of a truth table, one wants to make sure that no row duplicates any other, and that every possible row appears. A simple mathematical calculation tells us how many rows are needed; to check for duplicates, however, one goes through an extremely elementary, but absolutely necessary, piece of reasoning. One checks the first row against every other row, and then sets it aside to be used in no more comparisons. The same thing is done with the second row, and so on. This process involves truth-functional reasoning, and without it one could not divide the cases up or even be sure that the task was finished.[54]

Finally, depending on the algorithm, simple geometric or topological facts may be needed as well. For example, the application of an algorithm may involve dividing space on a piece of paper into boxes of a certain size, and geometric or topological reasoning about these boxes may be called for.

Let me also note parenthetically, although I will not make a lot out of it in this book, that the class of geometrical, logical, and arithmetic truths required here is also required in other evidential contexts, such as in confirmation procedures used in the sciences generally and in our daily lives (e.g., when using ordinary hardware tools such as rulers).[55]

I tentatively call the truths so required in the fashion we have described *evidentially central.*

Four important points before turning to the epistemic implications of these observations: First, notice how gerrymandered the class of

54 Of course this is the long way to do it. There are obvious shortcuts available, but they involve elementary logical reasoning too.
55 See Putnam 1975d, pp. 295–7.

truths involved here is from the point of view of the mathematical sciences. The logical truths required are quite limited (nothing, to be rather generous, with more than thirty connectives, or five quantifiers, is needed). Similarly, the mathematics used in applying such algorithms is elementary and possibly finitistic.[56] What is required is strictly limited in scope; a necessary condition is that one be able to do it in one's head.

Second, what I have said here about one needing certain mathematical truths in order to apply algorithms may seem problematic. If we need certain mathematical results to apply algorithms, how do we get *those*? Certainly I can't argue they are hardwired into us, for there are societies where the capacity for counting, say, is absent. Well, it seems clear that certain mathematical results are available to us without the application of any algorithms at all. For example, '1 + 1 = 2' seems to be available this way. It might seem otherwise, because when it comes to many formal systems, that result is a derived one. But, in practice, nobody learns it that way. One just memorizes it. Also, some algorithms are applied without relying on any truths at all. We simply have dispositions to apply them, and learning in some cases is no more than building up dispositions so that we can apply them by rote.[57]

It is worth noting that, in principle, any algorithm we apply could be *entirely* mechanized, and therefore applied *without any reasoning*, be it mathematical or logical (this, in fact, is why it is an *algorithm*). But we do not learn to apply any of the algorithmic procedures that we learn that way, and in fact, we couldn't because of the sheer fatigue that it would engender in us. (This is why, for example, designing Turing machines to do various tasks is so arduous: We can't expect them to reason out *any* aspects of what they have to do.)

Another equally crucial indicator that we do not apply algorithms in an entirely mechanized way is that, unlike computers, we make mistakes regularly – cannot help making mistakes regularly – and so must check over our work. By the way, in checking over the application of an algorithm, often more reasoning is used than in the orig-

---

56 Notice I said 'applying', and not 'justifying'. I stress again that to justify an algorithm often proves to be mathematically ambitious, 'justify' meaning here, of course, *proving formally* that the algorithm does what we take it to do.

  The claim that the mathematics needed to execute algorithms seem to be finitistic may seem to be false because notions of continuity underlying spatial idioms are traditionally handled in an infinitistic manner. But this has to do with a certain program: the attempt to anchor geometrical truths in an arithmetic foundation. As we shall see, there is pretty much no connection between the epistemological status of a mathematical or logical truth as far as its function in evidence gathering is concerned and its eventual location in standard mathematics.

57 But doesn't *this* answer raise problems with the purported normativity of the knowledge such dispositions are supposed to embody? Well, maybe. See Section 6.

inal application of it. For example, in checking a truth table, I often utilize more reasoning than I did in actually constructing it. This is because I spontaneously introduce shortcuts to lighten the workload. In general, it varies from individual to individual, and even from time to time for the same individual, what aspects in the application of an algorithm are mechanized (applied by rote) and which ones aren't.[58]

The third point is that, although I started out by noting that there seemed to be certain *truths* needed to apply algorithms, and then apparently went on to give a characterization of those truths, I think what was shown was quite a bit less than that. What we need to carry out these algorithms is *whatever* will license the steps we take when invoking these truths. Consider again the process of counting 'T's and 'F's. What is needed for *this* is nothing so grand as arithmetical truths, say, that '1 + 1 = 2'. In general, as far as this algorithm is concerned, adding 1 to 1 could fail to get us 2 almost all the time, provided that it didn't fail when counting 'T's and 'F's.

So, rather than saying that certain truths are needed, I will merely say that what is required is a certain sentential pattern with positive truth conditions assigned to the atomic principles (and where the list of positive truth conditions required is quite small). This pattern is contained in, for example, standard logic and mathematics, as traditionally interpreted and applied.

This third qualification is very important because, as it will turn out, no standard mathematical or logical truth is actually required to execute algorithms; but the explanation for the intuition of incorrigible co-empiricalness turns on the fact that the principles in the sentential pattern with atomic truth conditions that are required are regularly indicated (in our conceptual scheme, anyway) by such truths. Therefore, I claim that it is these habits of regular application of certain standard logical and mathematical truths, and the (somewhat) inchoate recognition that the class of principles indicated by these truths are necessary to these applications, that breed the impression of obviousness that those truths possess.

In any case, I should speak now only of those sentential patterns, with the truth conditions assigned to their atomic principles, as

---

58 Some philosophers have thought that Turing machines make good models of how the mind works. It may seem that I am implicitly denying this possibility by drawing a distinction between what is done "by rote" and what is done "by reasoning." But surely no argument has appeared in what I have written that shows that what we do "by reasoning" is not Turing computable.

I don't want to give the impression that I have legislated any position on this issue at all. I am making no claims about what is cognitively going on when someone "reasons." My point is only the epistemic one that, for example, there are (at least) two ways one can construct a truth table. The first way is entirely mechanical and can be executed by someone with no knowledge. The second way involves counting (at least) and does presuppose numerical knowledge (at least); and it is impossible for humans to carry out all their algorithms via the first method.

required for carrying out algorithms (and empirical evidence gathering) and, therefore, as *evidentially central*. What seems to be evidentially central, that is, is a set of principles.

Fourth, and finally, I have still overstated what has been shown by the considerations developed here. It is not the case, I think, that *every* principle appearing in the evidentially central sentential pattern is *required* by our algorithmic practices. Adapting a point I illustrated by means of Descartes' views a while back to our current terminology, we should distinguish between those principles that *are* used to execute algorithms and those principles that *must* be used to execute algorithms. Now it seems to me that there are no particular principles that must be used to execute algorithms. That is, we could consider modifications in the evidentially central sentential pattern achieved by "punching holes" in it. As long as *most of it* remains intact, it seems to me that our capacity to execute algorithms may be crippled but will not collapse altogether. On the other hand, I have already argued that it is not possible for us to eliminate *all* such principles from our algorithmic practices.

This qualification is vague. How much can be removed? I really can't say. But I can make the observation that pragmatic constraints – and, after all, what we need or don't need by way of principles for executing algorithms is a pragmatic issue – are generally sloppy. They rarely require something specific, for one can usually get around the requirement by using something else. Remove too much, however, and there is trouble.

So, finally, it is not even a sentential pattern of principles that is evidentially central but only a class of such patterns. The patterns in this class overlap in most of their principles, but have, I believe, a null intersection.[59] I will continue to call the set of principles in the union of this class of patterns, *evidentially central* with the implicit proviso that it is not any particular one of these principles that is required but only most of them (numerically speaking).

The reader may feel that I have not exactly described the possibilities here. It is not, the protest may go, that one should be considering "how much can be removed" but, rather, whether something *different* can be substituted. And, although I may be convincing on the claim that it is impossible to eliminate all principles from our algorithmic practices, it has not been shown that is it impossible to *replace*, step by step if necessary, all the principles involved with different ones.

Now I think there are two constraints here, and together they rule out the possibility contemplated as pragmatically incoherent. The first is that the alternatives have to be genuine alternatives; they can-

---

59 I stress again that the principles are being individuated, more or less successfully, by their location in the sentential pattern and by the truth conditions associated with their atomic constituents. This is sloppy, but there isn't much I can do about it.

not be principles we are already familiar with in terminological disguise. The second constraint, and this is a relative argument, is that the new principles must be able to underwrite the evidential procedures we already have. Although this is a relative constraint, it is still fairly strong. The reason is that the evidential procedures I have in mind are not merely ones that enable us to prove theorems of standard logic and mathematics; they are also used for theorem proving in the unapplied alternatives. This means that when we contemplate changing standard logic and mathematics to an alternative that has been worked out in some detail (and how are we to change to an alternative that has not been worked out in some detail?), we must be able to use the alternative to underwrite what we have done to establish its own theorems in the first place. But this means, at least as far as the principles used to execute algorithms are concerned, that it can only differ from them terminologically, or, to put it another way: For the most part, the sentential pattern of the principles we use must be contained in the sentential pattern exemplified by the alternative we are contemplating.

It was natural, in describing how the sentential pattern in question is involved in carrying out algorithms, and in discussing the alternative sentential patterns possible for doing this, to treat logic and mathematics together. But now, in order to discuss the epistemic implications of the discussion so far, it is natural to treat them separately. So we treat the logical principles first and turn to the mathematical principles after.[60]

Those sentential principles, which are logical and are needed for algorithms, seem only partially regimented in formal logic. By this I mean that it seems that a large number of the evidentially central principles do not translate successfully into the standard predicate calculus. For example, and I admit the contention is arguable, talk about propositional attitudes, counterfactual talk, higher-order reasoning, and talk of truth *simpliciter* do not seem to be so captured. Of course Quinean projects of regimentation attempt to show that such idioms, if recalcitrant, are *dispensable*. But this is controversial and, I suspect, false. As an example, propositional-attitude talk cannot be excised because, first, the execution of algorithms is at root a group effort and one needs to reason about the beliefs of others, and, second, at present (and, I think, for good) we have no alternative idiom that can do the job we need propositional-attitude talk for. Of course, one can also attempt to enrich the formal tools at one's disposal be-

---

60 This is a crude distinction, made primarily in terms of how the principles are generally regimented; that is, whether as logical truths or mathematical truths. Nothing particularly significant rides on sustaining this distinction with respect to the class of evidentially central principles.

yond the standard predicate calculus, and this is an active and ongoing set of projects.

But however such principles are to be modeled formally, it is that class of inference patterns that we use in our day-to-day execution of mathematical and logical algorithms that is evidentially central: In denying the validity of (most of) *those* patterns of inference, we would be undercutting the evidential basis of the mathematical sciences themselves. Consequently, regardless of how we view the status of a formal language with respect to a natural language,[61] a necessary condition for any formal system we finally adopt is that it leave *most* of our evidentially central inference patterns veridical.

And indeed, for those inference patterns in natural languages most tractable in terms of the standard predicate calculus, this is exactly what has happened. Large chunks of logic texts are taken up with the issues of translation. Inference patterns in natural language are either justified by a successful translation into the standard predicate calculus, set aside as problems for further work because the patterns are crucial for our day-to-day practices (e.g., counterfactual talk), or explained away pragmatically both by pointing out that the inference pattern is not crucial to what we do and by suggesting what kind of systematic mistake people make when they carry out such an inference (e.g., inferring 'Some people are happy' from 'All people are happy').[62]

Notice what has not been argued for here. Even given a syntactically characterized class of sentences, and given a deductive relation defined on that class, there are generally many semantic interpretations compatible with this relation.[63] But our situation is worse: The class of evidentially central inference patterns is small from a syntactic point of view, and its members are often linked to specific applications. Thus, and this is the crucial point, such inference patterns are not prima facie wide in scope. For the same reason, there is no obvious way to read off of the inference patterns underlying our evidence procedures anything as strong as the claim that, say, standard logic is classical, or even that some version of the deduction theorem (e.g., Γ, $A \vdash B$ iff $\Gamma \vdash (A \rightarrow B)$) must hold of it. Thus it is *always* open to us to honor the evidential centrality of such inference patterns, while in-

---

61 One might take the view that there is no fact of the matter about what logic natural languages have. Regimentation, as Quine views it, is a project with this perception of natural languages. On the other hand, one may think that there certainly is a fact of the matter about what the logical form(s) of a sentence in a natural language is (are).

62 Indeed translation from natural language into formal systems has all the markings of scientific practice generally when faced with recalcitrant data.

63 Such a fact is behind certain kinds of independence proofs. See, for example, Church 1956, pp. 112–8.

troducing new nonlogical nomenclature and denying the applicability of our old logical inference patterns to them; that is, it is *always* open to us to deny the apparent *scope* of a particular inference pattern.[64]

None of this is to deny that general considerations of simplicity and other theoretic virtues may be used to narrow the logical possibilities available for "rounding out" the set of evidentially central inference patterns into a full interpreted logical system. But my concern is with the actual inference patterns used to carry out algorithms, and how much (or how little) they restrain our options in choosing a co-empirical logical system.

Indeed, the restriction is a slight one, for, as we have seen, it has no impact on an assignment of truth values to the (infinitely many) logical truths beyond a certain size. It has no impact on whether we choose a classical logic or an intuitionistic one; indeed, because it forces no restrictions on the scope of the logical principles, there is very little by way of restraint on applications of this set of logical inferences to new situations.

We turn now to the principles of elementary mathematics. First off, as I have taken for granted, it is likely that retaining the evidentially central mathematical principles is compatible with strict finitism. If so, such principles don't even require mathematics beyond the capacity of first-order logic, for the mathematics needed is easily definable logically.[65]

But suppose this is false. Suppose good arguments can be given for the claim that the principles that license our execution of algorithms require the number sequence to be open-ended. Even so, Benacerraf's problem,[66] the fact that mathematical objects such as numbers may be identified with anything that can be given the appropriate structural properties, indicates how little restraint our counting practices place on what the standard mathematics used to license these practices must look like.

Geometry, however, may seem to be a different matter. If our principles, however meager, are Euclidean principles, then they are all false – space is not Euclidean.

But to take this view is to presuppose that the standard mathematics that the evidentially central geometric principles are eventually to

---

64 For example, we could introduce quantum logic this way. It would apply to the pristine terms of physics (some of them anyway), while leaving our old inference patterns intact where we need them (in applications to macro-objects). Of course, if we were good physicalists, we would have to explain how the use of such (stronger) inference patterns was justified in the special case (of macro-predicates). See later where I illustrate this move in the case of geometry.

65 With a fixed finite upper limit on the size of numbers needed, one can first-order define statements such as, 'The number of G's is the same as the number of F's'.

66 As described in Benacerraf 1965 and Part II, Section 8.

be embedded in must take such principles at face value. But, in fact, the inferences licensed by these truths are left intact even in a post-Newtonian universe: We *still* apply Euclidean truths – and not just in the context of evidence gathering, as it turns out.[67] The inferential *pattern* with its atomic truth conditions has been left intact. What has changed is that a story must be told to explain *why* so much is left intact. Here is the story: Spatial phenomena in our neighborhood successfully approximate Euclidean space because the violating curvature is below our perceptual threshold. Indeed, this fact is explicit in what kind of curvature we attribute to space when we take it to be non-Euclidean – whatever it is in the large, it still better be true that what goes on spatially in our neighborhood approximately fits what can go in Euclidean space.

Let me stress again that mathematical principles, as they are indicated by terminology used in ordinary unregimented language, do not reveal in what way their semantic status will eventually be anchored in standard mathematics – that is, whether they will be taken more or less literally or whether they will be treated as special cases. This is how they can have enough content to do the work we have them do while simultaneously not supplying enough constraints to tell us how the inference patterns they license should be embedded as co-empirical truths in our evolving conceptual scheme.

By way of closing this section, I will tie up a few loose ends and summarize what I think I have accomplished.

First, I should say a few words about the relationship between the intuitive epistemic status of the evidentially central logical and mathematical principles and their place in standard logic and mathematics. Generally, how crucial such principles are to executing algorithms is *not* reflected by their place in formalized mathematics. For example, the number-theoretic results, the geometric truths, and the simple logical inferences I have discussed are all on a par evidentially speaking. That is, they are psychologically accessible to pretty much the same degree, they are equally crucial for the cranking out of algorithms needed to derive mathematical and logical truths, and they are equally necessary to gather evidence in the empirical sciences.

Nevertheless, after such truths find their place in formal mathematics, broader ontological and epistemological issues tend to obscure their unregimented properties. For example, the fact that first-order logic is complete whereas set theory (and arithmetic) is not tends to give first-order logic epistemic priority over mathematics. Similarly, from an ontological point of view, the reduction of geometrical objects to (set-theoretic constructs of) arithmetic ones has tended to

---

67 Much of applied physics uses Euclidean geometry rather than the more complicated relativistic alternative.

align geometry with higher-order infinities, despite the fact that such notions are entirely alien to the geometric truths utilized when, for example, cutting tile.

The focus on systematicity makes it easy to overlook how much *doesn't* change when one contemplates alternative conceptual schemes. For example, if we switch logics, this will change (relative to empirical application) the topic-neutral inferences. And from the point of view of pure logic, the changes will be striking. But the actual inferences previously licensed may be largely untouched. What will have changed, of course, is the systematic *justification* of such inferences; but these considerations should not be allowed to obscure the substantial continuities that would be left intact in our inferential practices.

Recall the opening remarks of this section. I noted Parsons's point that Quinean considerations of centrality cannot explain the apparent obviousness of (certain) mathematical and logical truths. We now have the resources both to explain that obviousness and to explain why the Quinean apparatus had no hope of doing this job. These truths which seem obvious are precisely the ones we use to license the principles that are evidentially central, and it is their role as such that explains their obviousness. This is in part a psychological fact: They are used regularly. It is also in part an epistemic fact: The inferential pattern they exemplify in this application is (for the most part) ineliminable. The Quinean apparatus has no resources for explaining this obviousness because it approaches mathematical and logical truths *only* after they have been formalized. And it considers them only as they are applied as a whole to the empirical sciences. Thus the theoretical centrality of such truths is easily forthcoming but not the epistemic properties of that gerrymandered chunk of principles nearest and dearest to us that underlie the most obvious of such truths.

§5 Why There Are No Incorrigible Co-empirical Truths

Although the intuitions of obviousness (regarding certain statements of logic and mathematics) that have been the focus of attention in the last section have proved troubling to Quine's views about the epistemic status of mathematical and logical truth, I know of no source for an objection to his views based on these intuitions until Parsons raised the issue in his 1979–80. When Quine, more than forty years ago, wrote of the "totality of our so-called knowledge or beliefs," and claimed that "no statement" therein "is immune to revision,"[68] philosophers were suspicious because they felt that what he claimed

68 Quine 1951a, pp. 42–3.

simply couldn't be true of certain statements. For example, Quine's claim seemed wrong of certain standard examples of analytic truths, such as 'All vixens are female foxes'. In a friendly amendment Quine applauded, Putnam (1962) exempted these examples and others like them, using his notion of one-criterion words. The amendment applied only narrowly, leaving the epistemic force of Quine's claim untouched.

But, I think, the doctrine kept its philosophical feel of being incredibly radical, despite this modification, because of the (apparent) implication of the thesis that the principle of *noncontradiction* is revisable;[69] I think this is philosophically more shocking than the potential loss of the analytic truths.[70] There is something, one thinks, so very basic about this principle that it is a sine qua non of rationality itself. This thought directly emerges in attempts to show that Quine's picture is made *incoherent* by applying his claim to the principle of noncontradiction.[71]

On the other hand, intuitions about the grip the principle of noncontradiction has on rationality are disturbed in turn by the existence of deviant logics in which strong forms of the principle of noncontradiction are false.[72] Worse, these deviant logics are not merely uninterpreted calculi; they are serious attempts at constructing self-referential languages, and conceivably could turn out to be the best way to carry out this project.

Putnam, in a series of important papers,[73] explores the relationship between the apriority of the principle of noncontradiction and rationality. Choosing not a strong version of the principle of noncontradiction, but the weakest: 'Not every statement is both true and false', a principle compatible with deviant logics, he gives what I think are the strongest arguments available for why this statement should be regarded as a priori true.

Now, I believe that the claim in question is, in my own terminology, this: *'Not every statement is both true and false' is incorrigibly co-empirical.* It is not merely that the statement in question is true, and we are in-

69 In this context, the word 'principle' is not functioning as a term of art, as I had it do in Section 4, but rather as part of a piece of standard nomenclature. One generally writes, '*principle* of noncontradiction', and I follow suit in this.

70 This, at least, was my experience whenever I heard philosophers express informal criticisms of the views expressed in Quine 1951a. One also sees the same intuitions at work in (some) objections to the claim that logical truths are the results of convention. Of course this is not so of Quine's objections to the latter claim.

71 For example, Katz 1979.

72 See, for example, Priest 1979 and the *Journal of Philosophical Logic* 13, no. 2 (May 1984), for other articles.

73 I have in mind Putnam 1983d, 1983e, and 1983f. These articles depict several "flip-flops" in position. Putnam's explicit reason for exposing his vacillation in print is the "metaphilosophical" one that philosophers rarely expose their indecision this way. But he does eventually settle down on the claim that there are, indeed, what he calls absolutely a priori truths.

corrigibly sure of this; the claim is that the statement is true of our own conceptual scheme.[74] It is this thesis that I take Quine to have denied in Quine 1951a, and Putnam to be defending, and it is also the thesis that many philosophers find the denial of simply bizarre.

Before stating my views on this, let me get a couple of terminological issues out of the way. In the rest of this section I will generally not use the term 'incorrigibly co-empirical', but will try to stick to terms like 'a priori true' or 'absolutely a priori true', because I will be evaluating certain textual arguments of Putnam's rather closely, and I would like to stick to his terminology as much as possible. But I have made clear in my own terms what I think the claim amounts to, and the reader should keep that in mind both when evaluating what I think the thesis I am denying comes to and when evaluating whether I am indeed addressing these intuitions on their own grounds. I will, however, use the term 'co-empirical conceptual scheme', and mean by this 'conceptual scheme' in Quine's sense. As we have seen from Part II, this is only part of our conceptual scheme on my view, for it does not include alternate logical and mathematical systems that are not applied empirically.[75]

Now here is my claim: Despite the informal impressions of many philosophers, Putnam's attempts to turn these impressions into arguments and even some further backsliding on the part of Quine,[76] I think Quine's original view is the right one: No statement is a priori true.

But, in claiming that Quine is right here, I am not claiming that, broadly speaking, it is possible to revise the truth value of any statement in the co-empirical conceptual scheme. Rather, as we will see, for some statements, revision must take the form of being "revised out" of the co-empirical conceptual scheme altogether – that is, being *inexpressible* there.[77]

Up until this point I will have been defending Quine against Putnam's amendments. But, ironically, in so doing I will not have exactly

---

74 Recall the discussion in Section 3 where I distinguish metamathematics as a purely mathematical discipline and metamathematics as an empirical discipline. The intuition here I take to be the substantial, and co-empirical, one that the statement in question incorrigibly holds of *our* conceptual scheme.

75 I trust the use of the term 'co-empirical conceptual *scheme*' will not mislead. I mean, actually, 'subscheme', but the term is horrendous enough already. I also hope that the use of the term 'a priori' here will not be confused with the use of it I made in Sections 2 and 3.

76 During the question period after a paper Quine gave at Tufts in the fall of 1991, Quine seemed to suggest that he accepts Putnam's claim about the weak form of the principle of noncontradiction. He also seemed to suggest that he now regards this entire issue as philosophically uninteresting.

77 Two caveats: First, it may seem that Davidson 1986c, using broadly Quinean principles, has ruled out the coherence of the kind of revision I describe. On the contrary, but I cannot discuss the matter further now.

preserved Quine's epistemic insights. Traditionally, one role it seems a priori truths were supposed to have was that of being stability points in a conceptual scheme: They were supposed to be truths, come what may, one could rely on. But the (a priori) logical truths were supposed to serve a second (and distinguishable) purpose: They were to be descriptions of ways in which conceptual schemes would not change; they were to be markers of the limits of epistemic possibility,[78] the borders beyond which rational believers feared to tread. In dropping a priori truths, therefore, Quine apparently eliminated constraints on conceptual change – anything (nearly enough) becomes epistemically possible.[79]

Perhaps, therefore, it is no surprise both that Quine has found himself with "little to say of [the scientific method] that was not pretty common knowledge" (1986c, p. 493) and that principles of charity loom so large in his methodology and in that of fellow travelers such as Davidson. Descriptions of rational practice emptily become only the description of our own practices.[80]

Contrary to this, the result of this section will be that the two roles of a priori truths, in the traditional sense described earlier, come apart, and consequently the rejection of a priori truths does not simultaneously empty epistemic possibility of content: Although there are properties that *every* rationally accessible conceptual scheme must

Second, in saying that such statements have been revised out of the co-empirical conceptual scheme so that they are not even expressible there, I am not claiming that such statements are inexpressible in the conceptual scheme altogether; for they may be quite present in one or another alternative logic. Such a logic is part of the conceptual scheme, although terminologically restricted from being applied in the co-empirical part of that scheme.

78 I understand *rationality* (and this is vague, unfortunately) as the set of epistemic practices one *should* have. We picturesquely capture rationality "extensionally" via the notion of an *epistemically possible conceptual scheme*. This class of schemes is identified from the vantage point of the conceptual scheme we currently hold. A caveat: In modal logic, generally, one distinguishes between the entire set of possible worlds, and that set of possible worlds *accessible from* the actual world. A similar distinction is pertinent here. Because we are concerned with "what we can and cannot be wrong about," and not general considerations about epistemic logic, we consider only those epistemically possible worlds accessible from ours. I call them the *rationally accessible conceptual schemes*. A notion of epistemic possibility rather close to mine may be found in Putnam 1983f.

79 This was seen as an implication of Quine's claims pretty early on. Grice and Strawson 1956, write: "The point of substance (or one of them) that Quine is making, by this emphasis on revisability, is that there is no absolute necessity about the adoption or use of any conceptual scheme whatever" (p. 211). But Grice and Strawson do not see this claim as incompatible with the existence of analytic truths, where they gloss the latter (somewhat as Carnap does) as necessities *within* conceptual schemes. I interpret such "necessities" as truths, which, like mathematical truths, have evidence procedures establishing them that are independent of the empirical sciences. Any analytic truths learned in the process of learning a language will be so learned.

80 Putnam (1983g, p. 245) claims, "[T]he 'normative' becomes for Quine ... the search for methods that yield verdicts that one oneself would accept."

have, and such properties can be described (indeed, I have attempted to do just that in Section 4), they are not described by a priori truths. Neatly put, not every rationally accessible conceptual scheme has the resources to describe what is required in any rationally accessible conceptual scheme.

We start, then, by noting that there are reasons to be suspicious of the claim that the truth value of *any* statement is up for grabs, at least when it comes to logic. In the *locus classicus* of the position (Quine 1951a), Quine devotes most of his time on analyticity, and, indeed, most of the subsequent literature has followed suit by worrying about his objections to *that* notion. When Quine finally applies his revisability claim to logic, he merely invokes quantum logic as an alternative to classical logic. But that, or intuitionism (another respectable alternative), seems to have rather a lot in common with classical logic. Our earlier discussion in Section 4 (see especially note 53) suggests that it is problematic for us to describe what is held in common here as particular statements. But, if we semantically ascend to descriptions of the logics from above, it may be easier to capture what intuitively seems to be similar among them. Indeed, Putnam (1983b) seems to attempt just this when he fastens upon the statement 'Not every statement is both true and false', as an example of an a priori truth. Certainly, intuitively, it seems that a statement like this is one that we can take to hold of conceptual schemes based on either classical logic, quantum logic, or intuitionistic logic and one that is expressible in all three conceptual schemes based on such logics.

There is an important qualification that must be made here, however, although it is not an easy one to explain. Consider the project of describing the logical laws used to police the totality of our (co-empirical) knowledge. Pertinent to this project is an indeterminate collection of semantic and syntactic "concepts." Given a language, we can have at our disposal all the concepts necessary to do the job if we describe the language *from outside*, from the vantage point of a metalanguage, as it is commonly put. Doing so, however, leaves the principles of the metalanguage uncharacterized. On the other hand, there are strategies on the market for doing the job *from within* the vantage point of the language under study. But on pain of paradox such approaches must leave one or another aspect of the object language uncharacterizable.[81] As we shall see shortly, this fact bears di-

---

81 The literature in this area is enormous. See my citations in Part II, Section 3, note 14. For an explicit discussion of the impossibility of a "universal" language, see Herzberger 1980–1.

I should stress that the truth predicate is often blamed for the threat of paradox. But, contrary to this impression, there is a global problem here that may be solved by tinkering with the truth predicate, or the logic, or the capacity of the language to describe its own syntax. At present it is simply wide open which strategy or combination of strategies is the best way to go.

rectly on the question of whether there are a priori truths.

Let us turn, therefore, to Putnam's claim that

(*) Not every statement is both true and false

is an a priori truth. As I read Putnam 1983b, he starts by assuming that any theory that negates (*) will have to be a theory that consists of every statement and its negation. Putnam sees such a theory as empirically useless.[82]

So far so good. But couldn't we adopt (*) while restricting the application of universal instantiation to it? Putnam's response is that in doing so, we are playing verbal games: we simply don't mean what is normally meant by 'every statement is both true and false.'[83]

In his "Note" to 1983e, Putnam recognizes that his reliance here on (some sort of) theory of meaning allows us to define two ways in which a statement can be revised. As before, we can simply negate some statements (or change their truth values). But others can be revised by challenging the concepts they contain. Putnam suggests that intuitionism exemplifies this latter move: It replaces the classical notions of truth and falsity with something else. Thus Putnam now finds himself upholding a (slightly modified) version of the Quinean thesis that every statement is open to revision.

But in his "Note to supersede (supplement?) the preceding note," Putnam reverses his stand again. For even if we concede that certain concepts have been rejected, that doesn't show that no statement is a priori. Consider the statement, 'If the classical notions of truth and falsity do not have to be given up, then not every statement is both true and false'. That statement is a priori true regardless of whether or not we reject the classical concepts of truth and falsity.

This won't do, and there are two ways to see this. First the quick and dirty explanation: The hypothetical *mentions* the notion of truth in the antecedent and *uses* it in the consequent. But the use of a notion is verboten in a context where we have rejected it (a recipe for embarrassment: Assert the hypothetical in an intuitionistic context and watch what happens). We may be able to *describe* the classical notion of truth in a co-empirical conceptual scheme from which the concept is absent, but that doesn't imply that we can *use* it (or a satisfactory translation of it). This is not just a technical error either;

82 Why? Not for any particularly *deep* reason. Such a theory makes no distinctions at all; it is *pragmatically* useless to us. This is not a property peculiar to theories that are syntactically inconsistent. A theory that has *no* implications (imagine that the only inference rule admitted is the null one) will be equally useless. By the way, it is not obvious that such theories *are* empirically useless. See later.

I should add that Putnam clearly intends this statement to be one that is both *about* the language it describes and is *in* the language it describes. Furthermore, there are principles of disquotation being presupposed that enable us to derive from it every sentence and its negation in the very theory it is couched in.

83 Putnam 1983e, p. 102.

there is no way to do what Putnam needs the hypothetical for unless we retain the notion so rejected.

Here is the deep reason behind the failure just cited. *Truth* really is a problematic notion.[84] On some formal models of its operation, one simply cannot use the notion to say things like, 'Not every sentence is both true and false', *as Putnam and other philosophers – myself included – intend its interpretation*, either because there *is* no single truth predicate with sufficient scope (e.g., on Tarski's approach) or because quantification is restricted (e.g., on Parsons's approach). Other solutions that allow *something like* the above remark to be made are restricted in other ways.

In general, the characterization of the logical laws that hold in a co-empirical conceptual scheme calls for some capacity to describe the language the laws apply to. One cannot simply list the logical truths but must mark out the desirable classes of sentences syntactically, utilizing one proof procedure or another. In cases where the class of sentences is semantically complete with respect to one or another semantics, we can do so also in terms of truth.[85] But when self-reference is involved, the "descriptive capacity" available cannot be too rich on pain of paradox, and it is precisely which way is best to limit such capacity that is in question: For all we know, part of the best solution will cost us our capacity to express (*).

In Putnam 1983f, I detect a somewhat different argument. Consider the *Absolutely Inconsistent Rule (AIR): Infer every statement from every premise and from every set of premises, including the empty set.* Here are Putnam's claims:

(1) The AIR cannot be accepted by any rational being.
(2) A fully rational being should see and be able to express the fact that the AIR is incorrect.

Two qualifications: (2) is not meant to commit us to the a priori status of (*), but rather to the a priori status of the statement best expressed in *our* co-empirical conceptual scheme by (*). That is, we are not a priori committed to the logical form of (*), because the statement expressed by (*) may be expressible in an alternative co-empirical conceptual scheme where quantification (for example) is absent. (2) thus means that *some* statement in the co-empirical conceptual scheme of said rational being should translate as a rejection of the AIR. *Our* most reasonable candidate is (*), but that is not an a priori requirement.

Second, (2) is not meant to imply that any rational being whatever should be able to express beliefs about the *classical* notion of truth.

84 See note 81.
85 For a detailed illustration in the classical case, see Quine 1970, chapter 4.

Rather, Putnam claims that (*) is *generic*, that is, it does not rely on a particular notion of either 'truth' or 'statement'. The notions it uses are pretheoretic (and presumably compatible with a wide array of theoretic elucidations: intuitionistic, classical, epistemic, whatever).

Certainly, claiming that the notions of 'true' and 'statement' employed in (*) are pretheoretic detours around the roadblock that disabled Putnam's hypothetical. But, alas, it doesn't really come to grips with what I called "the deep reason" for the failure of the hypothetical. Putnam points out, quite correctly, that an objection to generic notions on the grounds that they are not regimented runs afoul of the fact that *most* of our working language is not regimented. This is true: We rely on unregimented notions all the time, and because, presently, no one approach to the truth predicate is recognized as right, our ordinary use of 'true' is clearly unregimented also.

But this doesn't give us carte blanche in our employment of 'true'. In particular, when issues arise about what *requirements* rationality places on co-empirical conceptual schemes, we cannot assume that it must be that "a rational being should see and be able to express the fact that the AIR is incorrect." For suitable regimentations of the notion of truth may exclude just that very ability.

Unregimented truth clearly gives us (or *some* of us, anyhow) the impression that natural languages are universal. But this impression cannot be used to supply a requirement on rationality: not with paradox looming around the corner. Notice this is the case *even if* we decide that the best regimentation of *our* co-empirical conceptual scheme does allow the expression of the unacceptability of the AIR.[86] For, currently, whatever logic we take as the best candidate for us does not exclude substituting something else later. More is called for than being *our* best choice if a principle is to be a priori.

Intuitions that illegitimately rely on the purported universality of the language our co-empirical conceptual scheme is couched in are not the only ones that contribute to the sensation that there is something a priori about principles of noncontradiction. For example, there are strong intuitions that it is not possible for us to *discover* that a statement and its contradictory are true, for example, 'John is running' and 'it is not the case that John is running'. Considering specific examples like this gives the impression that I pulled a sleight of hand when I invoked the rather theoretical universality considerations ear-

---

86 Does the Tarskian approach allow "a fully rational being to see and be able to express the fact that the AIR is incorrect"? Superficially, at least, it doesn't seem to. At any stage in the hierarchy, one's truth predicates only range over a portion of the statements pertinent. But perhaps a schematic understanding of how to iterate the Tarskian hierarchy suffices for "being able to see" the truth of (*) – although I must glumly point out that there are technical problems with iterating said hierarchy. Being able to *express* the truth of (*), however, is a different matter.

lier and applied them to notions, such as 'truth', which we use to state *laws* of logic. For the irrationality of contradiction, one might argue, lies not in the mere assertion of a law of logic that allows contradictory instances, but in the instances themselves allowed by that law. What experiences could one possibly have that would even *tempt* one to think that both a statement and its contradictory should be inferred?

The objection just raised suggests that although we might not be in a position to say explicitly that certain statements are a priori true (because our language lacked resources to talk about its own properties), they would be for all that. We could recognize intuitively that certain statements could not be false even if our language lacked the capacity to state such a fact about those statements.

But I think even this goes too far, for it seems that it is easy to manufacture examples where one is tempted to think that both a statement and its negation *are* true. Imagine that we have a rather well-developed theory about a certain area. The theory predicts one thing, say, but observation yields the opposite; and so we are tempted to retain *both* claims.

Well, maybe the example is not as convincing as I suggested (temptations are notoriously matters of taste). We usually don't settle for such contradictions – we tinker with theory and/or methods of observation until they no longer arise; and I have no argument with this. But we *could* decide that, in fact, certain pairs of contradictory statements are true, and avoid vacuity by simply restricting our inference rules from operating on such instances, as I will show at the end of this section. Our not doing so thus seems to be a top-down methodological consideration, *not* a matter of our failing to see how it is possible for us to be tempted to think that both a statement and its negation are true.

But the objection might arise: Surely we couldn't *observe* a contradiction, could we? We couldn't observe, upon opening a box, that a sheet of paper is red and that same sheet is not red – in fact, we know a priori that such a thing couldn't happen, right? And this is apart from the general desire to avoid contradictions because, say, they make our inference schemes overrich.

Alas, invoking unimaginability here is simply a red herring. Certainly I can't *imagine* seeing that a sheet of paper is red and that same piece of paper is not red, but I can't imagine seeing a sound either. Nevertheless, it isn't a priori that sounds can't be colors. One can imagine theory changing in such a way that we come to claim that certain colors *are* sounds. I don't mean to suggest that something unimaginable would then happen (i.e., that, *au* Churchland, we would then come to *see* sounds), but certainly we can't take the mere unimaginability of something to indicate its impossibility. Theory is more subtle than that. (A slogan: What is unimaginable is not necessarily inconceivable.) This means that if we take the fact that we can't imagine observing contradictions as indications that they are impossible, that is already the application of a methodological decision to possible experience. It really isn't *further* evidence that such intuitions indicate the presence of something a priori.

Well, maybe it is possible for us to imagine coming to believe that certain statements and their negations are true, but could we come to believe that *every* statement and its negation are true? Putnam writes:

We can imagine all of the predictions of non-Euclidean physics coming true, even if we happen to be Euclidean physicists. But we don't know what it would be like for all the predictions of the theory that consists of every statement together with its negation to come true. (1983e, p. 103)

However, if my arguments about our intuitions regarding particular contradictions are right, I don't see what the problem is supposed to be. Certainly we could come to believe that a particular contradiction holds, and if we chose not to modify the inference rules *as far as the semantics of our language was concerned* but only as far as pragmatics was concerned, then we would think that every statement and its negation were true (and false, for that matter). I don't particularly care for the idea, but I don't think it is all that incoherent either. On such a view (it has mystical overtones, I guess) we would be quietistic about what we could really know (or say, for that matter). The view would be that our belief that we can say something (or know something) to the exclusion of something else would be based on pragmatic and ad hoc modifications in our modes of inference. One would be humble in the face of the fact that our apparent acts of communication were delusional in a way barely expressible (e.g., 'every statement is true and false'). As I said, the possibility has nothing to recommend it other than the fact that it is a possibility.[87]

A caveat:[88] I stated that invoking unimaginability is a red herring. But there is more packed into that argument than appears in my exposition, and I should make it perfectly clear exactly what is involved here. In arguing that the inability to imagine an alternative to a statement's truth is *never* a criterion of its incorrigibility, I supported myself with the consideration that we cannot tell in such a case if our failure of imagination is due to there not being such a possibility or

87 Notice that the principles (discussed in Section 4) that we use in evidence gathering are underwritten in this case by *pragmatic* rules rather than semantic ones.
88 The next six paragraphs are inspired, in part, by a conversation with Jerrold J. Katz.

due to a mere psychological failing. But notice a premise here: I am presupposing that there is a verificationist requirement on incorrigibility: Any distinctions relevant to our knowledge claims must be ones that are epistemically accessible to us.

Let me illustrate this by means of Descartes' *cogito*. By meditating on this statement (indexed to me), I draw the conclusion that it cannot be false. But then, one might wonder, is it that it cannot be false, or is it that I simply have had a failure of imagination? Indeed, it could be either that I am wrong about the statement (although it seems impossible for me to think that *I* might be wrong about the statement) or that in fact the statement doesn't even have the semantic properties that I think it has.[89]

Notice that if one rejects the verificationist premise here, one has a rejoinder. There is a distinction between cases where a mere failure of imagination is involved and cases of actual impossibilities (regardless of whether one is capable of drawing this distinction in practice), and one simply has incorrigible knowledge in a case where one has hit upon an actual impossibility even if, in fact one *cannot* prove that a mere failure of imagination is *not* involved.

For example, suppose that the *cogito* has the semantic properties that Descartes takes it to have, even though he cannot rule out the possibility that it does not (in particular, it really is expressible in the language Descartes is speaking); then it cannot be false.

Now I repeat that it seems to me that a requirement on genuine incorrigibility is that *all* the distinctions relevant to the epistemic status of the knower's claim *be available* to the knower. In this case, even if it is a fact that the *cogito* has the semantic properties that Descartes takes it to have, he cannot guarantee this fact to himself, and so his knowledge of the *cogito* is not incorrigible knowledge.

Notice that my acceptance of the verificationist premise for incorrigibility is not a *requirement* that the notion of incorrigibility must support a verificationist premise (indeed, it had better not be, on pain of contradiction). So, certainly, one can consider redefining the term 'incorrigible' in a way that denies that premise. I have to say that I have no interest in terminological wrangles. If someone wants to rework the term 'incorrigible knowledge' so that it is possible to have incorrigible knowledge without having access to all the facts relevant to the epistemic status of that knowledge, he or she is welcome to do so. Indeed, it certainly is possible that we might want to so change what we take 'incorrigible knowledge' to mean. But I have not been concerned with a notion so modified, and for a good reason: I have

---

89 This is not so crazy, at least not if what I have suggested about 'truth' earlier is not. After all, the suggestion there was that our commonsense intuitions about the semantic properties of the idiom 'true' are wrong.

been concerned with whether the distinctions a knower can make are sufficient to guarantee him or her that he or she is right about a statement's perceived status, and for that the modified notion of incorrigible knowledge (lately contemplated) is not pertinent.[90]

I promised to offer a simple method of changing the truth value of *any* sentence without (very much) impairing the utility of the conceptual scheme it is in. Here is the method:

(1) Reverse the sentence's truth value.
(2) Rewrite the inference rules so that none of them applies to the sentence.

A caveat: Clearly this method is generalizable to any syntactically characterizable class of sentences. If the class of sentences is small enough, the co-empirical conceptual scheme is only minimally impaired (and one cannot rule out an overall gain in simplicity, given changes elsewhere in the co-empirical scheme). That is, the resulting scheme is rationally accessible. On the other hand, change the truth values of *enough* sentences and the rational accessibility of the result cannot be guaranteed.

Another caveat: In pressing this example, one must not confuse a co-empirical conceptual scheme with a particular *postulate basis* used to generate (some of the) sentences in that scheme.[91] A postulate basis may have a fixed number of *sentences* as premises and one or two inference rules (including, possibly, substitution). Looked at this way, the suggestion that the truth value of *any* sentence is up for grabs may seem absurd, as it could be a postulate in the postulate basis (and cost us too many sentences). But what must be always considered is what the total class of sentences being excluded is. Generally speaking, we can use any number of equivalent postulate bases to describe the same set of sentences in a co-empirical conceptual scheme, and care must be taken not to let the parochial character of a particular postulate basis fool us into overestimating the centrality of a particular instance of a logical truth merely because it is central to that postulate basis.

Yet another caveat: Whether we regard the propositions expressed by the sentences as outright *excluded* from expression in our co-

90 Here's an objection: Don't we all know (well, *most* of us anyway) that verificationism is a bad thing? Well, sure. It is a *bad thing for semantics.* One doesn't, for example, want to tie notions of semantics to practices of verification. One doesn't want criteria of meaningfulness to be verificational ones. (As I said, most of us, these days anyway, feel this way.) But this is because one doesn't want one's *epistemology* messing around with one's *semantics.* And this shows that verificationism is a *good thing* when it comes to epistemology. By the way, I am not willing, necessarily, to argue that verificationist criteria apply to notions of *justification.* I am only arguing that one should take such criteria seriously when concerned with *incorrigibility.*
91 See Part II, Section 2.

empirical conceptual scheme will depend on whether we can find a substitute that "expresses the same statement."

One final caveat: In claiming that this method yields rationally accessible co-empirical conceptual schemes when applied to sufficiently small sets of sentences, I am not obliged to actually describe situations where the optimal scheme would be one of this sort. But it is worth pointing out that the *one thing* that prevents this sort of method as a solution to liar's paradoxes in self-referential languages is that the class of viciously self-referential sentences is not syntactically characterizable.

§6 Normative Considerations, the Success of Applied Mathematics, Concluding Thoughts

We have considered several sorts of apriority that mathematical practice can exemplify and have accepted some and rejected others. As I mentioned at the outset of Part III, my purpose here has been to explore the various ways in which elements in the practice of mathematics are epistemically special when contrasted with the empirical sciences.

One way that was discussed fairly thoroughly in Part II, but has not yet been brought up in this context is the identification of systems by stipulation (see Part II, Sections 6 and 7). There is a sense in which such stipulations are a priori because although they can be motivated by applications to the empirical sciences, in no way does their justification lie in the results of these applications. Notice that this sense of 'a priori' does not naturally apply to sentences or epistemic warrants, for its impact is found, rather, in its creative effect on the statements we can take to hold of a particular subject matter. Intuitively, there is some analogy with certain earlier notions of the a priori,[92] but it would be unnatural, I think, to try to recast what is going on here in terms, for example, of epistemic warrants for sentences.

I mention this aspect of mathematical practice again, however, not just for the sake of completeness but because it bears on the topic of normativity, to which we now turn.

Recall that one issue here is to explain the gap between a person's mathematical practices and what are taken to be the correct practices. The possibility of mistakes requires a distinction between these things, and we go on to find sources for such a distinction and, consequently, for the possibility of error.

Mathematicians can make several sorts of mistakes, and they do not all have the same status. First off, let us consider a simple sort of

---

92 For example, see Kripke's (1980) famous discussion of the meter stick.

case: the computation of sums (in a particular fixed terminology). My claim is simple. There is an algorithm here that has been adopted by the community. In saying this, I am making a descriptive point. The fact that such an algorithm has been adopted enables us to predict, ahead of time, how the community will resolve mistakes (provided they stay with that algorithm), and how they will identify who is good at the practice (of computing sums) and who is bad.

Now the crucial thing that confers objectivity on the algorithm is that it is a mechanical procedure, and the results that follow, given this procedure, can also be predicted ahead of time. As I observed in Section 4, however, the fact that there is a mechanical procedure for doing something does not, in general, imply that the procedure will be carried out in a mechanical way. Indeed, I have argued that this is rarely how algorithms *are* carried out, especially because mathematicians use a semiformal terminology that can be very deceptive. Consequently, the results in practice can often deviate from what the procedure would result in if it were carried out mechanically. But, I claim, the existence of such a procedure acts as an objective constraint for correcting errors.

I think that this sort of thing holds wherever one is faced with a single system that has been in place in a community of practitioners for some time. But often, as we have seen, systems are replaced by other systems. In such a case, the question of whether a "mistake" has been made is far less robust, far more of a projection, than in the first case. From one point of view, what we are really facing in such cases are alternative possibilities – and sometimes the mathematical community is acutely aware of this.[93] But other times the modifications in the systems used are so gradual and so continuous that it becomes very hard to judge whether one is faced with actual errors within a single system, a shift in systems, or both.[94] Often contemporaries, or near descendants, decide the issue in terms of the system, or set of systems, that eventually carry the day. The early practices can then be misinterpreted by a reading back of the later system into them.

My point here is not to object to the practice of calling derived results that hold of earlier systems, but not of later ones, *mistakes*. This is a perfectly acceptable practice. For in calling something a mistake, the community of mathematicians is repudiating it. My point is only that mistakes come in degrees, and the degree of a mistake, what we might informally call its epistemic value, turns on the extent to which

93 For example, this is how Quine (1969a) describes the options in set theory, and this view was certainly shared at one time.
94 This is largely how I see the evolution of analysis in the seventeenth and eighteenth centuries.

the mistake depends on a thick ontological commitment, as opposed to a thin one.[95] Crudely, the epistemic value of a mistake metaphysically depends on how much it involves getting something wrong about the world. To the extent that the mistake is due to a shift in the stipulations at work among possible systems, it is epistemically lightweight.

These considerations about errors I regard as deep, at least in the sense that it leaves a great deal of detailed work undone: I have barely touched the question of when, historically, systems used by mathematicians change, and when, in fact, mathematicians are simply making errors. Nor do I think this detailed work at all easy to do. But I am not enough of a scholar to shed much light here and content myself with merely pointing out the general framework.

But there is another problem of fairly ancient vintage that involves normativity as well. We might call it *the objection to Mill*.

Here is a (probably inaccurate) depiction of Mill's view. Mathematical truths are established empirically because mathematical objects are empirical. Either they are properties of empirical objects or they are particular empirical objects. Now, the objection goes, this view runs afoul of the normativity of mathematics. We have deep-seated intuitions that in fact even if certain statements we take to be true of mathematical objects failed to hold of the objects or properties designated as mathematical objects, this would *not* show that such statements were false. Rather, it would show that such properties and objects were *not* the mathematical objects we took them for.

Now the question is: *Why* is this the case? What is it that is special about mathematical objects, or the laws that they obey, that seems to make them immune to being ontologically reduced to physical objects (of one sort or another)?[96] The answer, I am afraid, is that this is not a genuine problem. Given what mathematical objects are used for (as referents for noun phrases within and across systems), and given how mathematical truths are established (via the twin operations of algorithm and stipulation), there simply is no need for an on-

---

95 A caveat: I think that if a mistake is due to a deviation from an algorithm that is generally accepted, this involves something that is comparatively ontologically thick – namely, certain dispositions on the part of the community. Unfortunately, I will not give an argument for this claim, for any such argument must come to grips with the rule-following problem, and I cannot discuss that problem in this book.

96 Historically, the normativity of mathematics (or at least logic) in this sense is blended with its irrefutability. For example, consider Frege (1967, p. 14) where he writes: "But what if beings were even found whose laws of thought flatly contradicted ours and therefore frequently led to contrary results even in practice? The psychological logician could only acknowledge the fact and say simply: those laws hold for them, these laws hold for us. I should say: we have here a hitherto unknown type of madness." That aspect of these intuitions based on the apparent irrefutability of mathematics (or logic) is not being discussed now because it has already been dealt with in Sections 4 and 5.

tological reduction of this sort. And, in fact, were we to succeed in carrying one or another such reduction out, the properties of the things the mathematical objects would be reduced to could only get in the way.

Historically, the apparent normativity of mathematics and logic seemed to mark them out as a special species of knowledge, one that might call for a fairly radical metaphysical explanation. We found something different. We have found that this sort of normativity has its source in the fact that we like to change the systems we use without changing the apparent ontology. It also has its source in the fact that how we use mathematics does not require its truths to be directly anchored referentially to the world. Applied mathematics must, when conspiring with empirical science, result in an empirically viable unit. But this places no requirement on what mathematical objects are, and so what they *are* is not dictated by applied mathematics.

The upshot is that the normativity of mathematical truth has turned out to be an indicator not of something philosophically rich, something deep that we must provide an explanation for. Rather, it has turned out to indicate something *shallow* – mathematical objects in a sense that we have seen are less, both metaphysically and epistemologically, than their empirical brethren.

We turn now to the success of applied mathematics. Daston (1988, p. 221) begins a recent article with the following remark: "By almost all current philosophical accounts, the success of applied mathematics is a perpetual miracle." Whether or not such a broad swipe at so many philosophical accounts is sensible, one hears in any case an echo of the Kantian question about how pure and applied mathematics are possible.[97] No book-long discussion on the philosophy of mathematics is complete without at least an indication of where an explanation can be found for the success of applied mathematics.

Consider the following analogy. Even a cursory observation of nature seems to reveal a surprisingly good fit between organisms and their environments. Until the advent of Darwin's theory of evolution, there seemed to be only the unpleasant option of invoking a divine hand in design to explain this.[98] But once a theory of evolution is in place, one has an explanation, although one that is not so much a

---

97 Friedman (1988, p. 82) says Carnap was "concerned above all" with this question.
98 Two caveats: First, a lot of the debate over design swirled around the question of providing an explanation of how organisms "so complex" could have arisen without divine guidance. But, because of the vagueness of the notions of 'complexity' and 'divinity', I regard this as an uninteresting version of the problem. The point, I think, has to do not with complexity or divinity, but fitness and design. Second, I really believe (but clearly cannot argue for it now) that to the extent that an explanation of the form "someone built these things" could be given (and certainly a story like that which did not include Christian or even religious trappings was logically possible), it was the only respectable hypothesis available until the theory of evolution emerged.

story of the survival of the fittest as a story of the demise of the unfit. Two factors are crucial to the story told here. First, there is the extinction of those who are not successful – those, that is, who do not fit very well in the niche they are born into or subsequently trapped in. And second, there is the point that survival is not a matter of a *perfect* fit between organism and niche but only one good enough to enable the organism to get by long enough to reproduce.[99]

Our story about the success of mathematics is similar. We have the advantage that Kant did not enjoy of being aware of how much alternative unapplied mathematics is available and possible. We are also aware that it is only in modern times, relatively speaking, that the fact that mathematics can be generated independently of possible applications has been fully exploited by mathematicians. Often, as I have indicated, alternative mathematical theories were not developed simply because mathematicians had implicit applications so clearly in mind.[100]

Notice, further, that it is not mathematics alone that meets the tribunal of experience, but, as the nod to Quine's phrasing indicates, the vast corporation of applied mathematical theory and empirical scientific theory that so meets experience. Furthermore, this vast corporation does not apply without (substantial) glitches. It is not that one finds a smooth fit between possible experience and the mathematics and empirical theory that we take to explain what we experience. Rather, it works well enough to get by with, but not well enough to avoid anomalies altogether.[101] And in resolving such anomalies, empirical theory may change, but the mathematical theory applied can change as well. The story is not a fresh one any longer, but it works for all that.

But one may still feel I have not been fair to the opposition. Look, the objection may go, let's run with your evolutionary analogy a little further. Surely someone can protest that even if the evolutionary story, as you've so far given it, explains how complex organisms can evolve to (more or less) fit niches that they occupy, *it still doesn't explain* how life is possible in the first place. For that one needs some explanation of how, *in the first place,* creatures could arise that evolve in a way that is sensitive to selection. After all, *most* stuff in the universe doesn't.

And, the opposition could continue relentlessly, the fact that there is an answer to this question shows that the question is a meaningful one: The sort of answer I have in mind is one that invokes self-

99 This is part of why the popular tendency to treat 'natural' as synonymous with 'healthy' is so comical.
100 Recall my discussion of arithmetic in Part II, Section 6.
101 On the ubiquitous presence of anomalies, see, among others, Lakatos 1976b.

replicating molecules, or, more significantly, self-replicating molecules that don't *quite* self-replicate. Once this notion is in place, we have a sense *both* of how life can evolve and how it can arise in the first place, even if many details in the story are missing.

Now what is required for applied mathematics is something similar. One may have an evolutionary story that explains how applied mathematics can evolve once it is clear that mathematics can be (more or less) applied in the first place, but we still lack an explanation of how any mathematics can be applied at all.

In order to help motivate my response to this objection, let us first consider how the traditional platonic story attempts to handle this problem. Mathematical objects, in that story, are not in space or time; and one accesses mathematical truths, recall, by accessing such objects. But how does a believer in this story explain the success of applied mathematics? Well, adopting Plato's line, one can try to connect mathematical truth to empirical application by claiming that empirical objects, in some sense, are *copies* of mathematical ones.

This move is not likely to be appealing to contemporary realists, especially because the fact that empirical objects are copies, in some sense, of mathematical objects is an unexplained brute fact on this view. But notice that, traditionally, what is appealing about realism in science is precisely that the objects one takes to exist are supposed to explain the success of science, and this explanation is missing for the mathematical realist who does not subscribe to something like Plato's copy theory.

Now, clearly, it would be ironic if the view of mathematics that I developed with the intention of leaving the mathematical object out of any explanation of how we come to know pure mathematics subsequently required the mathematical object to explain the success of mathematics. So it is clear that the view I intend to argue for, although one that the traditional platonist can adopt, is not one that she will find supportive of a position which requires that the mathematical object have a role in explaining the success of applied mathematics.

But is there *any* sort of story that I can tell? Well, here's one. Consider various sorts of objects that one can *count*. How is this possible? The answer is that the objects have certain properties that make them susceptible to being counted. For example, they are discrete, tend to exist for (relatively speaking) long periods of time, can be distinguished easily, and behave decently when it comes to principles of individuation. Not all objects are like this, as anyone who has faced a bowl of jellyfish will tell you.

Now the interesting thing is that, to some extent, the susceptibility of many objects to being counted is *not* an unexplained brute fact about those objects. That is, it may follow from scientific facts about their properties that they are susceptible to being counted. (Consider how we might see how it follows from the physical structure of certain objects that they are discrete, tend to exist for, relatively speaking, long periods of time, etc.) Note the analogy to self-replicating molecules. If our chemistry and physics are primitive enough, *that* there are self-replicating molecules may be an unexplained brute fact. That is, we may be unable to explain why there are such molecules or how they work. On the other hand, we may be able to explain this, and therefore, we may be able to explain, in terms of chemistry and physics, why there are objects to which evolutionary theory can be applied.

The mathematical case, I think, is entirely similar. In certain cases we have an underlying explanation for why the mathematics that can be applied to the area can be applied. But sometimes this will be simply a brute fact.

Compare physics for a moment. At certain stages in the development of the field, certain physical facts are brute facts. For example, one may simply have to accept that the gravitational constant is the magnitude that it is. Later, however, one may be able to derive this from other physical laws. Similarly, that certain mathematical laws work when applied to an area may be, at a time, simply a brute fact. Later, one may find, however, that one can actually provide an explanation for why the mathematics applies.

Notice that the sort of explanation available for why mathematics (and logic) applies empirically has turned out to be similar to the sorts of explanations available for why empirical science applies empirically. And it should not be surprising that this happened. Consider logic for a moment. Currently, we like the idea of one uniform logic that applies across all empirical disciplines. But this isn't required. We might decide that the best way of doing science requires specialized logics. In this case there really would be no such thing as *logic* as we understand it. Rather, there would be broad principles limited to particular fields of study. But the explanation, to the extent we have one, for why such principles work would be *identical* to the sort of explanation for why, say, laws of motion work. That is, either it would be a brute fact or it would be a fact we could derive from other, more fundamental, principles.

Someone, having reached this point in the book, and more or less understanding what I have been about here, may still wonder: Is the position platonist or nominalist? I started the book by saying that I would defend something I called linguistic realism. But what relationship does that view have to the traditional platonist/nominalist dichotomy?

Let us distinguish three possible claims that a platonist may make, claims I draw explicitly from the introduction to Benacerraf and Putnam 1983.[102]

(a) Statements couched in the language of mathematics may have truth values regardless of whether the rules we use to generate mathematical truths determine what those truth values are.

Observation: Suppose first-order PA is our sole method for generating truths of arithmetic. Despite the incompleteness of PA, there is a fact of the matter about the truth value of every sentence in the language of PA. It is because he accepts the classical version of this claim, for example, that Quine has described himself as a realist, and it is this claim (also construed classically) that Dummett uses to characterize the realist position.[103] Although both take realism to be the acceptance of bivalence, I have generalized the claim. What seems essential, especially in Quine's discussion, is not that *every* sentence be assigned a truth value but that sentences can have truth values regardless of whether we have the means, even in principle, to assign such values to them. This could be the case even if one adopted a nonclassical logic.

(b) The noun phrases in the language of mathematics refer. For example, '1' refers to 1, '{ }', refers to { }, and so on.

Observation: Depending on how 'refer' is to be understood, this claim may or may not be stronger than (a). For example, one may think that the use of a Tarskian metalanguage suffices to characterize one's notion of 'refer'. In this case (the Quinean case), (a) is arguably (metaphysically) no stronger than (b). But one may feel that stronger conditions must be placed on the notion of reference before one is willing to say that a term refers. Arguably, this is how Field 1972 and Devitt 1981 should be understood.

(c) The objects of mathematics are abstract objects that are neither in space nor time and are acausal.

Observation: This third claim is easily divided into several more (just as the second claim was, actually), and one may find some of the properties mentioned more central or significant than others. Traditionally, one or another version of (c) has been seen as the most salient aspect of the platonist position; however, recent philosophers have come to regard either (a) or (b) as more significant, so that, for example, someone like Maddy (1990) can regard herself as a platonist despite her wholesale rejection of (c).

---

102 I should add that (1) I don't claim everybody who considers himself or herself a platonist subscribes to all of these claims, and (2) I don't claim that I have stated all the beliefs that a platonist may take to be central to his or her platonism. But I do think Benacerraf and Putnam have described the core beliefs of many platonists.

103 See Quine 1981b and Dummett 1963.

Now, because these three claims can be teased apart, and because they, and their various versions, can be held somewhat independently of each other, and also because it seems to me that there are many criteria for ontological commitment that are not always operating at the same time, with the same sorts of objects (the Quinean move of searching for one particular criterion seems to me deeply misguided), and because it seems hard to decide how to evaluate these criteria against each other (it isn't even clear to me that it makes sense to try), I guess I really regard the question of what a platonist is (and consequently, what a nominalist is), and whether I am one, as terminological questions *at best*. By way of sketchily summarizing my position, let me indicate why I think this.

We start with (a) and (b). There is a sense, surely, in which I accept both of these. For example, it seems perfectly reasonable, on my view, to adopt classical logic and its accompanying Tarskian semantics in contexts where there are *in principle* no ways of verifying the truth values of every sentence in the mathematical languages so adopted. Furthermore, if one prefers formalizing in a first-order way, one finds that my platonism, if we may call it that, is perfectly compatible with Quinean requirements on existential commitment. But I may even be regarded as a platonist in a stronger sense than Quine is. For I have stressed the co-referentiality of mathematical terms across theories, and even across the divide of what he might regard as different conceptual schemes. This way, as I have mentioned, mathematical posits as exemplified in the noun phrases used by mathematicians take on a robustness rivaled by their empirical counterparts (because the latter also often survive transplantation to new contexts). Any nominalist who has felt that denying platonism is tantamount to making drastic changes in one's semantics is likely, therefore, to regard the position sketched out in this book as platonistic.[104]

On the other hand, dyed-in-the-wool platonists are bound to feel cheated. To start with, in no sense, one may argue, does the position sketched here accept anything like (c). Furthermore, the platonist may complain, my "acceptance" of (a) and (b) is something of a cheat too. For in accepting, say, Tarskian semantics and the language of classical mathematics, and in taking terms to co-refer across systems, I am really not accepting them in the sense that I am claiming there are actually objects "out there" that mathematical terms refer to. For example, co-referentiality is not explained in terms of there being a

104 Notice, also, that because I accept classical mathematics I like (in that context, anyway) impredicative definitions. Some philosophers of mathematics regard such a taste as a mark of platonism – although I must confess I don't quite see how to evaluate my position using this sort of criterion, because I *also* accept nonclassical mathematics, in particular, mathematics wherein impredicative definitions are rejected.

particular object that two such terms refer to, but really just in terms of a stipulation that (metaphysically speaking) is arbitrary. In fact, one point of this book has been to explain standard mathematical practice in a way that *leaves out* the mathematical object. Isn't this tantamount to simply treating such objects as *not* existing while showing how we can continue to talk as if they do?

I am not particularly willing to argue with either way of describing what I have done here. Sometimes, as in the beginning of the book, where I was implicitly concerned with positions that call for the re-writing of ordinary mathematics one way or another to eliminate references to metaphysically inert mathematical objects, I think of myself as a platonist; for I am preserving traditional mathematical talk and practice, and in particular, the seamless identification of the semantics of mathematics with that of the empirical sciences, at least as far as talk of reference and truth is concerned.[105] Call this a *platonism without puzzles.*

On the other hand, if someone stresses the many ways in which I write as if there are no such mathematical objects at all – how I explain mathematical practice in ways that do not involve one's interaction with mathematical objects and how I explain mathematical truth in ways that do not involve getting something right about *something* – it is clear that I sound like a nominalist. But nominalism has always seemed to carry with it the requirement that those nominalistically inclined must adjust their language so that terms purporting to refer to the offending objects no longer appear; and such a process is invariably costly. I see none of this as required. Call the position, therefore (if you wish), *nominalism on the cheap.*

In any case I have not been overly concerned with the somewhat vulgar question of where the position I have sketched out belongs on the realism–irrealism continuum.[106] My purpose has been different;

---

105 Benacerraf 1973, p. 408 stresses the requirement that there be a homogeneous semantic theory in which the semantics for the propositions of mathematics parallels the semantics for the rest of the language, a requirement that he thinks perhaps reduces to the "plea that the semantical apparatus of mathematics be seen as part and parcel of that of the natural language in which it is done, and thus that whatever *semantical* account we are inclined to give of names or, more generally, of singular terms, predicates, and quantifiers in the mother tongue include those parts of the mother tongue which we classify as mathematese."

106 But here are some further thoughts on the question. Consider the categories given in Resnik 1980, p. 162: *Ontological platonism, epistemological platonism, realism,* and *methodological platonism.* I am an ontological platonist for I do not truck with attempts to reduce mathematical objects to physical or subjective entities. I am *not* an epistemological platonist, for I do not see the epistemology of mathematics and empirical science as on a par. I am a methodological platonist, at least insofar as standard mathematics is concerned, for I endorse nonconstructive mathematical methods (in nonconstructive mathematics, anyway – although in standard mathematics, too, it must be added). The hard question is whether I am a realist,

I have tried to take a careful look at mathematical practice and the intuitions that arise because of that practice, and see what these things show us about what we are doing epistemically and ontologically when we do mathematics. In so doing, my hope has been to show that certain traditional philosophical problems often raised about mathematical ontology or epistemology are not problems at all. But, more important, I have tried to expose what has been going on here in such a way as to quell the philosophical puzzlement that is invariably behind the posing of these problems.

By no means have I tried to "Wittgenstein away" the problems here; at least this is not what I have assumed I have been about. Certain problems I *have* regarded as misguided, and I have tried to show why. Other problems, problems that naturally arise from observing our mathematical practices and intuitions, I have tried to solve by giving a picture that explains what is going on. In doing this, I have always attempted to accommodate the intuitions I found; and although I have not *always* succeeded in accommodating them exactly at face value, I hope the position worked out in this book makes it clear what mathematicians are doing when they do mathematics, and what they are doing it *with*.

whether, that is, I believe the objects of mathematics exist independently of us and our mental lives. As with all hard questions, the answer is, "yes and no." The space of mathematical possibility is constituted by syntactic facts: what systems can exist and be connected. I am as realist about this as I am about empirical objects. However, mathematical objects *are* posits, and posits are not, strictly speaking, independent of their positors.

Appendix

The point of this appendix is to show that the truth predicate *really is* compatible with a wide array of logical systems. To make this point, we need to show that neither the classical logical setting nor the rich set-theoretic tools Tarski employs to define satisfaction are necessary for a truth predicate.[1] This is what we have set out to do here.

There is one other interesting point to make. Tarski, and due to his influence, other philosophers and logicians too, have generally used as a criterion for a successful theory of truth that Tarski biconditionals be derivable from it. I want to turn this procedure on its head. My theory will contain the Tarski biconditionals (as derivation rules) and some apparatus for referring to the sentences of the object-language (substitutional quantification). This will be taken as *sufficient* for a theory of truth. Anything further needed for a *semantic theory* of the object-language and for a description of the syntax of the object-language must be evaluated on other grounds. Interestingly enough, however, it will turn out that in certain contexts (the sentential calculus) what I give will be sufficient for a semantic theory too, but this will not generally be the case. Syntax is another matter altogether. We will see that descriptions of proof procedures for language call for resources that go beyond what we supply here.

I will not be concerned with self-referential contexts, which raise an entirely distinct set of problems. So our illustrations will involve an approach to the truth predicates that essentially respects an object-language/metalanguage distinction, in the sense that our new quantifiers are restricted in their range only over a part of the language we give.

---

1 Notice that the same remarks apply to the alternative approaches considered by Kripke, Herzberger, and others. For generally such approaches are all model-theoretic ones set in the classical context, and so they presuppose a set-theoretic ontology in addition to classical logic.

Tarski's metalanguages are far more powerful in their descriptive capacity than what we offer here. This is because they do far more than give a theory of truth for the languages they describe: They give a standard semantics for such languages, and they can describe the syntax of such languages. Such a capacity is important for certain purposes; but it simply isn't required for a truth predicate, for a device, that is, that enables us to assent to groups of sentences without explicitly listing them.[2]

We start with a logical system L in a language $\mathcal{L}$, both of fairly arbitrary character.[3] Our first task is to define languages $\mathcal{L}*$ and $\mathcal{L}**$ that will properly contain $\mathcal{L}$. Both $\mathcal{L}*$ and $\mathcal{L}**$ will contain the truth predicate for $\mathcal{L}$ and the logical systems couched in these languages will be the same; the languages will differ only in what terms they admit. In constructing these languages, I will sometimes be willing to use symbols from the underlying language $\mathcal{L}$ if available, but other symbols of $\mathcal{L}$, $\mathcal{L}*$, and $\mathcal{L}**$ must be kept clearly distinct. I indicate this explicitly. In the discussion to follow, the symbols of $\mathcal{L}$, $\mathcal{L}*$, and $\mathcal{L}**$ generally stand as metalinguistic names for themselves. Greek letters generally function as metalinguistic variables here, and where so used, implicit quasi-quote conventions are in force to uphold use/mention distinctions. Occasionally I utilize quote marks explicitly, and where I do so, I invariably employ: ", rather than: ', because the latter is a special symbol of $\mathcal{L}*$. Finally I will presuppose underlying syntactic notions such as *wff, sentence, constant,* and the like, in $\mathcal{L}$ when they are available there. In particular, certain languages, such as that of the first-order predicate calculus, presuppose a distinction between *wff* and *sentence,* although others, such as that of the sentential calculus, do not. In what follows, where I speak of *sentences,* I mean *sentences,* and not *wffs,* except when the distinction does not exist in the particular language $\mathcal{L}$. In such cases *wffs* are meant.

Languages

$\mathcal{L}*$ and $\mathcal{L}**$ have the following vocabulary: the vocabulary of $\mathcal{L}$; a quote mark (distinct from anything in the vocabulary of $\mathcal{L}$): '; quan-

2 In Azzouni 1991, I stressed how the approach I explored there separated the theory of truth for a language from the theory of the semantics of that language and the theory of the syntax of that language. As it turns out Belnap and Grover (1973) stressed this division in an important paper where they explored substitutional quantification into quotational contexts. They describe it this way: "We have demonstrated that the theory of truth is *separable from* and does not depend on a theory of truth conditions" (p. 272), where a theory of truth conditions just is a theory of the semantics and a theory of the syntax of a language.
   I recommend comparing that paper with what I do here.
3 L will be of a fairly arbitrary character as a *system.* But, generally, the particular set of rules we use to generate L will have to be chosen carefully. See below.

tifier symbols: ∃, ∀; parentheses (if $\mathscr{L}$ doesn't have any): (,); equality (if $\mathscr{L}$ doesn't have it): =; sentential variables (distinct from anything in the vocabulary of $\mathscr{L}$): p, q, r, s, . . . ; a one-place predicate constant (distinct from anything in the vocabulary of $\mathscr{L}$): T$_r$; predicate variables of various sorts (distinct from anything in the vocabulary of $\mathscr{L}$): P, Q, R, . . .

Next, we give the definition of *wff* in $\mathscr{L}^*$ by building up the notions of *constant, term,* and *wff* inductively in the standard fashion:

Definition AP1 The following are the formation rules of $\mathscr{L}^*$:

(FR1) If Σ is a formation rule of $\mathscr{L}$, then it is a formation rule of $\mathscr{L}^*$. The free occurrence of variables in any wff formed according to an application of this rule are all and only the free occurrences of variables in the components the rule was applied to. If the resulting wff is a wff of $\mathscr{L}$, then it has no free occurrences of variables.

(FR2) If σ is a sentence of $\mathscr{L}$, then 'σ' is a constant of $\mathscr{L}^*$. There are no free occurrences of variables in 'σ'.

(FR3) If σ is a sentential variable of $\mathscr{L}^*$, then 'σ' is a term of $\mathscr{L}^*$. σ occurs free in 'σ'.

(FR4) If 'σ' is a constant of $\mathscr{L}^*$, then 'σ' is a term of $\mathscr{L}^*$.

(FR5) If σ is a sentential variable of $\mathscr{L}^*$, then σ is a wff of $\mathscr{L}^*$. σ occurs free (in σ).

(FR6) If τ is a term of $\mathscr{L}^*$, the T$_r\tau$ is a wff of $\mathscr{L}^*$. The free occurrences of variables in T$_r\tau$ are all those that occur free in τ.

(FR7) If τ and χ are terms of $\mathscr{L}^*$, then $(\tau = \chi)$ is a wff of $\mathscr{L}^*$. The free occurrences of variables in $(\tau = \chi)$ are all and only the free occurrences of variables in τ and χ.

(FR8) If Ψ is an n-place predicate variable of $\mathscr{L}^*$ and $\tau_1, \ldots, \tau_n$ are terms of $\mathscr{L}^*$, then $\Psi\tau_1 \ldots \tau_n$ is a wff of $\mathscr{L}^*$. The free occurrences of variables in $\Psi\tau_1 \ldots \tau_n$ are all and only the free occurrences of variables in $\tau_1, \ldots, \tau_n$.

(FR9) If ς is a wff of $\mathscr{L}^*$, and σ is a sentential variable of $\mathscr{L}^*$, then $(\exists\sigma)\varsigma$ and $(\forall\sigma)\varsigma$ are wffs of $\mathscr{L}^*$. The free occurrences of variables in $(\exists\sigma)\varsigma$ and $(\forall\sigma)\varsigma$ are all and only those of ς if σ does not occur freely in ς. If it does, then the free occurrences of variables in $(\exists\sigma)\varsigma$ and $(\forall\sigma)\varsigma$ are all and only those in ς minus the occurrences of σ. (If σ occurs freely in ς, then we say the free occurrences of σ in ς are in the scope of the quantified σ (the token of σ in $(\exists\sigma)$ and $(\forall\sigma)$, respectively) or are bound by the quantified σ.)

(FR10) Nothing is a wff or a term of $\mathscr{L}^*$ unless it is so by virtue of clauses FR1-FR9.

Observations:

(i) Notice that the notion of *occurrence of free variable* defined here has only to do with the sentential variables of $\mathscr{L}^*$ proper. Wffs of the underlying language $\mathscr{L}$ may have free variables with respect to $\mathscr{L}$, of course.

(ii) Notice that constants are *not* a substituent class for the sentential variables. For constants are of the form 'ς', where ς is a sentence of $\mathscr{L}$, and the variable replaces ς, not 'ς'. This is crucial for what is forthcoming: It will enable us to present the Tarski biconditionals in a closed form (see observation v, example 2).

(iii) Notice that we have allowed terms that name sentences of $\mathscr{L}$ but *not* wffs of $\mathscr{L}$ that are not sentences. This restriction is not unavoidable: We don't need the capacity to refer to wffs that are not sentences if all we are concerned with is *truth*. But where one wants a greater capacity to describe the syntax and semantics of a language, and where there is a distinction between sentences and wffs (such as in the standard predicate calculus), then stronger referential powers are desirable. I say a little more about this later.

(iv) Suppose that $\mathscr{L}$ is English (we suppress punctuation, and for the sake of simplicity pretend that none of our notation, especially the single quote, occurs in English). Here are some wffs:

(1) T_r'Snow is white'
(2) $(T_r$'p' if and only if ('John is running' $=$ 'q'))
(3) $(\forall p)$ ((John is running unless T_r'p') if and only if q)
(4) $(\exists p)$ $(T_r$'p' only if ('p' $=$ 'r'))

Notice that (1) has no free variables, "p" and "q" occur free in (2), only "q" occurs free in (3), and only "r" occurs free in (4). Because the quantifier here is going to be understood substitutionally, one should have no qualms about quantifying into quotational contexts. Some contortions are necessary if one tries to read these sentences aloud in a way that sounds natural. This, however, is no objection to them.

(v) Finally, notice that (FR1) allows us to take advantage of the logical resources of $\mathscr{L}$. To illustrate this, imagine that $\mathscr{L}$ is the standard sentential calculus. Here are some examples of wffs (as sentential variables of the sentential calculus, we will use Russian capital letters so that the sentential variables of the sentential calculus can be distinguished from the sentential variables of $\mathscr{L}*$):

(1) $(\exists p)$ (('Ж' $=$ 'p') $\Leftrightarrow$ T_r'Ж')
(2) $(\forall p)$ $(T_r$'p' $\Leftrightarrow$ p)
(3) $(((Ж \vee Б \Rightarrow Ш)$ & ('Б' $=$ '(Ш & Ж)')) $\Rightarrow$ $(\exists p)T_r$'Ш')

Recall that although the quotation context can be quantified into, we cannot substitute a sentential variable (of $\mathscr{L}*$) for any arbitrary string of symbols of $\mathscr{L}$. We can do so only for the whole sentences quoted.

Definition AP2 The formation rules of $\mathcal{L}^{**}$ are the same as those $\mathcal{L}^{*}$ (*mutatis mutandis*) except that (FR2) is replaced with

(FR2**) If ς is a wff of $\mathcal{L}^{**}$ containing, as vocabulary from $\mathcal{L}^{**}$ that is not in $\mathcal{L}$, at most only sentential variables, then 'ς' is a term. Any sentential variable occurring free in ς occurs free in 'ς'.

Observation: The substitution of (FR2**) for (FR2) has one (significant) effect: It allows a more complicated sort of quantification into the terms of $\mathcal{L}^{**}$. The sentential variable doesn't merely replace the entire sentence of $\mathcal{L}$ named by the term, subsentential parts of it can be replaced too. As an illustration of this increased power, consider again the examples of observation (v) following the formation rules of $\mathcal{L}^{*}$, and notice that the following wff of $\mathcal{L}^{**}$ is not a wff of $\mathcal{L}^{*}$:

(1) (Ⅲ ⇒ (∃p) (∀q) ('(p ⇒ q)' = '(p ⇔ Ж)'))

Please note that a term like: '(p ⇔ Ж)' is not to be understood as the name of: (p ⇔ Ж), as: '(Ж ⇔ Ж)' is taken to be the name of: (Ж ⇔ Ж). Rather, it involves a substitutional variable, and is not functioning as a name *at all.*

Inference Rules

Before giving the inference rules of L*, we need the following definition:

Definition AP3 Let Ξ be the set of all the sentences of $\mathcal{L}$. Let ς be a wff of $\mathcal{L}^{*}$ or $\mathcal{L}^{**}$, and σ a sentential variable free in ς. Then ς(Ξ) is an infinite list of wffs such that each wff ξ is in ς(Ξ) iff there is a sentence μ in Ξ such that ξ was gotten from ς by substituting μ for σ.

Observation: The notation "ς(Ξ)" is defined notation to facilitate our "writing down" infinite lists of wffs. It is not part of the vocabulary of either $\mathcal{L}^{*}$ or $\mathcal{L}^{**}$.

Definition AP4 The following rules are called *the inference rules of L*.* The wffs before the "⊢" in a rule are in the *antecedent* of the rule, and where more than one wff appears in the antecedent separated by commas, these are described as being in different *clauses* of the antecedent. There is no significance to the order of the clauses in the antecedent. What follows the "⊢" in a rule will be described as the *consequent* of the rule. Each wff that appears in an inference rule comes with an implicit *line number* and an implicit list of line numbers called a *premise list*. Unless the rule specifies otherwise, the consequent of a rule inherits the union of the premise lists of the wffs in the antecedent as its premise list. (None of the rules given here will explicitly specify otherwise. Of course, whatever specifications occur in the rules of L that (I1) licenses are carried up into L*.)

(I1) If Λ and Γ are sets of wffs of $\mathscr{L}$ and $\Lambda \vdash_{\mathscr{L}} \Gamma$ is licensed by an inference rule of L, then

$$\Lambda \vdash \Gamma.$$

(I2) For any term τ, we have

$$\vdash (\tau = \tau).$$

(I3) If ς is a wff of L*, and ς^* is gotten from ς by replacing zero or more occurrences of a term τ at ς by a term τ^*, then

$$\varsigma, (\tau = \tau^*) \vdash \varsigma^*.$$

(I4) If τ and τ^* are distinct terms with no free variables, and ς is any wff, then

$$(\tau = \tau^*) \vdash \varsigma.$$

(I5) If 'ς' is a term, then

$$T_r'\varsigma' \vdash \varsigma$$

and

$$\varsigma \vdash T_r'\varsigma'.$$

(I6) If σ is any sentential variable, ς any wff and $\varsigma[\kappa]$ a wff differing from ς in that every free occurrence of σ has been replaced by κ, which is either a sentence of $\mathscr{L}$ or a sentential variable, and further if ς has the same number of free occurrences of variables as $\varsigma[\kappa]$ if κ is a sentential variable, we have

$$\varsigma[\kappa] \vdash (\exists\sigma)\varsigma.$$

(I7) If σ is any sentential variable, ς any wff, and $\varsigma[\kappa]$ a wff differing from ς in that every free occurrence of σ has been replaced by κ, a sentential variable, and further if ς has the same number of free occurrences of variables as $\varsigma[\kappa]$, and σ does not occur free in any of the wffs whose line numbers appear in the premise list of $(\exists\sigma)\varsigma$, we have

$$(\exists\sigma)\varsigma \vdash \varsigma[\kappa].$$

(I8) If σ, ς, and $\varsigma[\kappa]$ are as in (I6), and further if ς has the same number of free occurrences of variables as $\varsigma[\kappa]$ if κ is a sentential variable, then we have

$$(\forall\sigma)\varsigma \vdash \varsigma[\kappa].$$

(I9) If σ, ς, and ς[κ] are as in (I7), and σ does not occur free in any of the wffs whose line numbers appear in the premise list of ς[κ], then we have

$$\varsigma[\kappa] \vdash (\forall \sigma)\varsigma.$$

Finally, because we are employing substitutional quantification, we have an infinite derivation rule.

(I10) If σ and ς are as in definition AP3, then

$$\varsigma(\Xi) \vdash (\forall \sigma)\varsigma.$$

Observations:

(i) The rules are understood to be construction rules for derivations. That is, at any stage in a derivation, if each clause in the antecedent of the ⊢-rule appears at some line in the derivation, and the other constraints specified in the statement of the rule are satisfied, then what appears in the consequent may be added to the derivation at that stage.

(ii) Notice that the constants here do not act like the constants in the standard predicate calculus. In particular, just because we can deduce a sentence ς from the sentence of T,'ς' doesn't mean we can deduce ς from (∃p)T,'p'. On the contrary.

(iii) Notice that no rule corresponding to (I10) is available for the existential quantifier. This is because we are not employing a full Gentzen calculus (and giving an interpretation to an infinite list of sequents in a rule of inference). However, if classical negation is available, we will be able to deduce the existence of valid infinitary derivation rules that apply to the existential quantifier. (See Theorem AP1.)

(iv) Notice that the point of (I4) is that distinct terms with no free variables never refer to the same things.

Derivations of Infinite Length

The use of substitutional quantification introduces a complication not present with first-order objectual quantification; unless certain measures are taken, strong completeness is lost. One possible measure to preserve strong completeness (which was employed in my 1991) is to allow the introduction of fresh constants into the language.[4] Because we are treating systems here as fixed in their vocabulary, this is not an appealing solution. Thus we can preserve

4 See Dunn and Belnap 1968. Also see Leblanc 1983.

strong completeness by adopting the infinitary derivation rule (I10), and with it derivations of infinite length.

But if we just introduce arbitrary (well-founded) derivations of arbitrary (countable) ordinality, we will be faced with a derivation system that is not user-friendly. However, the mere admission of *certain* infinite derivations does not pose complications in manipulating such derivations if the class of admissible infinite derivations is tame enough; that is, if the sorts of infinite lists of sentences admissible in derivations is severely restricted.

Our results (at least as far as completeness is concerned) will be disappointing. We will introduce a user-friendly derivation system where (I10) can be used only finitely many times in a proof—but completeness will be shown with respect to unrestricted use of (I10). Whether sharper results are possible will not be addressed here.

We need a perspicuous notation for the infinite lists of sentences that will be admissible in derivations, which notation I give immediately below (it resembles the notation given in Definition AP3). Observe that the fact that such a perspicuous notation is even available for the set of infinite lists of wffs admissible in derivations is an indication of their tractability.

Definition AP5 Let ς, σ, and Ξ be as in Definition AP3, and let $\rho_1, \ldots, \rho_n$, where $1 \leq n < \infty$, be distinct sentences of $\mathcal{L}$. Then

(i) $\varsigma(\Xi)$ is as in Definition AP3.

(ii) $(\varsigma(\Xi)/\rho_1, \ldots, \rho_n)$ stands for the same list as $\varsigma(\Xi)$ *without*, however, those n wffs gotten from ς by replacing σ with $\rho_1, \ldots, \rho_n$, respectively. We call $\varsigma(\Xi)$ and $(\varsigma(\Xi)/\rho_1, \ldots, \rho_n)$ Ξ-*wffs*.

Definition AP6 We call the following rules Ξ-*rules*. We will talk of the "antecedents," "consequents," and "clauses" of these rules as in Definition AP4. Also, let ς, σ, Ξ, and $\rho_1, \ldots, \rho_i, \ldots, \rho_n$, where $1 \leq i \leq n < \infty$, be as in Definition AP5, and let $\varsigma[\rho_i]$, for any i, be that wff gotten by substituting ρ_i for all free occurrences of σ in ς.

(i) $(\varsigma(\Xi)/\rho_1, \ldots, \rho_n), \varsigma[\rho_i] \vdash (\varsigma(\Xi)/\rho_i, \ldots, \rho_{i-1}, \rho_{i+1}, \ldots, \rho_n)$,
  $(\varsigma)\Xi)/\rho_1, \ldots, \rho_{i-1}, \rho_{i+1}, \ldots, \rho_n) \vdash (\varsigma(\Xi)/\rho_1, \ldots, \rho_n)$, and
  $(\varsigma(\Xi)/\rho_1, \ldots, \rho_n) \vdash \varsigma[\rho_j]$, where $j \neq 1, \ldots, j \neq n$.

(ii) $(\varsigma(\Xi)/\rho_n), \varsigma[\rho_n] \vdash \varsigma(\Xi)$,
  $\varsigma(\Xi) \vdash (\varsigma(\Xi)/\rho_n)$, and
  $\varsigma(\Xi) \vdash \varsigma[\rho_n]$.

Observations:

(i) These are premise-deleting rules. If the wffs appearing in the antecedent of such a rule appear on any lines in a derivation above the wff appearing in the sequent of that rule, then the line numbers of the wffs appearing in the antecedent *may* be deleted from the premise list of the wff appearing in the sequent.

(ii) Notice that the notation for Ξ-*wffs* is predicative. That is, we have made no provisions for quantifying into such things or combining them with the connectives of $\mathscr{L}^*$ ($\mathscr{L}^{**}$). (This should be no surprise given the observation following Definition AP3.)

Next, using Definitions AP5 and AP6, we introduce a finite derivation notation to represent the admissible infinite derivations. And for this purpose, we treat (I10) as a derivation rule with a Ξ-wff in its antecedent.

Definition AP7 Let Σ be a finite numbered list of wffs that are either in $\mathscr{L}^*$ ($\mathscr{L}^{**}$) or are Ξ-wffs. We say that a wff ς in Σ obeys an I-rule or a Ξ-rule of the form $\Theta \vdash \alpha$ if

(i) ς is α.

(ii) Each wff in a clause of Θ appears on a line in the list somewhere above ς.

Definition AP8 An L*-derivation sketch is a finite numbered list of wffs, where to the right of each wff is written a finite list of numbers, and where

(i) Each wff is either a wff of $\mathscr{L}^*$ ($\mathscr{L}^{**}$) or a Ξ-wff.

(ii) Each wff obeys either an I-rule or a Ξ-rule or is a *premise*.

(iii) Any premise contains its line number and only its line number in its premise list. Every other wff in the derivation has a premise list containing the numbers of the premise lists from the antecedent wffs in the I-rule or Ξ-rule that it obeys (subject to whatever provisos the rule subjects the premise list to).

Observations:

(i) We are going to use L*-derivation sketches as transcriptions of the admissible L*-derivations. Notice the following. Infinite lists of wffs are manipulated as wholes, and no concern is shown with the ordering of the wffs within them. To facilitate this, we adopt a canonical way of listing every sentence of $\mathscr{L}$ that imposes a recursive ordering on any list of wffs $\varsigma(\Xi)$ admissible in a derivation; that ordering is taken to be preserved on lists like $(\varsigma(\Xi)/\rho_1, \ldots, \rho_n)$ too. Any particular canonical listing that does the job will be fine; we omit details.

Notice, also, that the particular infinite lists of wffs are restricted in type. They must be either generated from every sentence of $\mathscr{L}$ by substitution for a sentential variable in a wff of $\mathscr{L}^*$ (or $\mathscr{L}^{**}$) or they differ from such a list only in missing finitely many wffs. If we have the tools for manipulating a list containing every wff of $\mathscr{L}$, there further lists pose no difficulty.

We now use L*-derivation sketches to define the notion of an L*-derivation.

Definition AP9 An *L*-derivation* is any (possibly infinite) list of wffs that, using the canonical list of sentences of $\mathscr{L}$, and the notation of Definition AP5, is gotten from an L*-derivation sketch. In the rewriting process, any line containing a Ξ-wff numbered n is replaced by an infinite list of sentences numbered $n1, \ldots, nm, \ldots$.

Definition AP10 We say that a *sentence ς follows from* a set of sentences Δ if there is an L*-derivation in which ς appears on the last line of the derivation, and the premise list of that line lists sentences only drawn from Δ. We say ς is an L*-theorem if there is a derivation in which it appears on the last line and the premise list on that line is empty.

L* couched in $\mathscr{L}*$ is powerful enough for the truth predicate, despite the fact that it largely respects the context it is grafted onto. However, it is generally too weak for the purpose of *semantics*. What more is needed? In the case of the classical sentential calculus, it turns out, and we will illustrate this, very little: The slightly richer capacity to describe the syntax of the underlying language given by L* couched in $\mathscr{L}**$ suffices.

The Classical Sentential Calculus

We start by taking $\mathscr{L}$ to be the language of the classical sentential calculus and L to be the classical tautologies (we'll call them $\mathscr{L}_1$ and L_1 to distinguish them from later examples). Because we did not presuppose any logical connectives in the general design of $\mathscr{L}*$, and because of (I1), the strength of L_1* turns on the particular formulation of L_1, even though such formulations of L_1 are equivalent in their derivational strength when attention is restricted to L_1 alone. Consider our case. One characterization of L_1 is the simple rule

(R1) If ς is tautological, then

$$\vdash \varsigma.$$

But if L_1 is characterized this way, there will be a great deal that we will be unable to prove (because we don't have modus ponens); for example, the deduction theorem cannot be shown in L_1^*. Generally speaking, because the rules added to L_1 to get L_1^* are derivation rules, L_1 itself needs to be designed carefully. This is a general phenomenon that goes beyond the sorts of systems we are concerned with here. For example, one finds it if one attempts to graft derivation rules for the standard predicate quantifiers onto the sentential calculus. In what follows we offer a characterization of the sentential calculus that is amenable to being combined with the derivation rules we have given.

The Language of the Sentential Calculus

The vocabulary of $\mathscr{L}_1$ is the following: syntactically primitive *sentences:* $S_1, \ldots, S_n, \ldots$, parentheses: (,), connectives: $\Rightarrow, \neg$.

Observation: Standardly, one studies the sentential calculus on its own by utilizing sentential variables. But we are concerned with it in the context where it is the underlying logic of an interpreted language. So we include the primitive sentences of that language explicitly. Our example of the sentential calculus is just that, a *calculus* for deriving truth bearers from other truth bearers.

Definition AP11 The following are the formation rules for $\mathscr{L}_1$:
  ($\mathscr{L}_1$1) If ς is a primitive sentence, then it is a wff.
  ($\mathscr{L}_1$2) If ς and ρ are wffs, then ($\varsigma \Rightarrow \rho$) are $\neg\varsigma$ are wffs.
  ($\mathscr{L}_1$3) Something is a wff if and only if it is so by virtue of clauses ($\mathscr{L}_1$1) and ($\mathscr{L}_1$2).

Observations:
  (i) Notice that there is no distinction in this system between wffs and sentences.
  (ii) We introduce "$\wedge$," "$\vee$," "$\Leftrightarrow$," via definitions in the usual way.

Definition AP12 The following rules (adapted from Mates 1965) are called the *inference rules of L_1*. We use "antecedent," "consequent," and "clause," as in Definition AP4.
  ($I_1$1) If ς and ρ are wffs, then

$$\rho \vdash (\varsigma \Rightarrow \rho).$$

This is a premise-deleting rule. The premise numbers of the new line are all of those of the previous line, except (if desired) any that is the line number of a line on which ς appears.

($I_1$2) If ς is a wff, and Θ is a finite set of wffs that tautologically imply ς, then

$$\Theta \vdash \varsigma.$$

Definition AP13 An L_1-derivation is a finite $L_1{}^*$-derivation in which every wff of the derivation is a sentence of $\mathscr{L}_1$.

We next give a few of the definitions, theorems, and derived inference rules that are available in $L_1{}^*$.

Theorem AP1 Let ς be a wff in which the sentential variable σ occurs free, let ρ be a sentence of $\mathscr{L}_1$, and let $\varsigma[\rho]$ be that sentence gotten from ς by substituting ρ for σ. Then we can show the following derivation rule:

$$(\exists\sigma)\varsigma, (\neg\varsigma(\Xi)/\rho) \vdash \varsigma[\rho].$$

Proof Sketch Start with $\neg\varsigma(\Xi)$, and derive $(\forall\sigma)\neg\varsigma$ from it with (I10). Next, using ($I_1$1) and Definition AP6 (ii), discharge $\neg\varsigma[\rho]$ and $\neg\varsigma(\Xi)$ as premises so that we have: $(\neg\varsigma(\Xi)/\rho) \vdash (\neg\varsigma[\rho] \Rightarrow (\forall\sigma)\neg\varsigma)$. From this, it is an easy and standard derivation to get: $(\neg\varsigma(\Xi)/\rho) \vdash ((\exists\sigma)\varsigma \Rightarrow \varsigma[\rho])$, and from this, the desired result.

Theorem AP2 If a sentential variable, then

$$(\forall\sigma)\,(T_r\text{'}\sigma\text{'} \Leftrightarrow \sigma)$$

is a theorem.

Corollary If ω is a tautologically true sentence of L_1, then $T_r\text{'}\omega\text{'}$ is a theorem.

Theorem AP3 If σ and κ are sentential variables, then the following are theorems:
    (i) $(\forall\sigma)\,(\forall\kappa)\,((T_r\text{'}(\sigma \wedge \kappa)\text{'} \Leftrightarrow (T_r\text{'}\sigma\text{'} \wedge T_r\text{'}\kappa\text{'}))$
    (ii) $(\forall\sigma)\,(\forall\kappa)\,((T_r\text{'}(\sigma \vee \kappa)\text{'} \Leftrightarrow T_r\text{'}\sigma\text{'} \vee T_r\text{'}\kappa\text{'}))$
    (iii) $(\forall\sigma)\,(\forall\kappa)\,((T_r\text{'}(\sigma \Leftrightarrow \kappa)\text{'} \Leftrightarrow (T_r\text{'}\sigma\text{'} \Leftrightarrow T_r\text{'}\kappa\text{'}))$
    (iv) $(\forall\sigma)\,(\forall\kappa)\,((T_r\text{'}(\sigma \Leftrightarrow \kappa)\text{'} \Leftrightarrow (T_r\text{'}\sigma\text{'} \Leftrightarrow T_r\text{'}\kappa\text{'}))$
    (v) $(\forall\sigma)\,(T_r\text{'}\neg\sigma\text{'} \Leftrightarrow \neg T_r\text{'}\sigma\text{'})$

Theorem AP4 If σ and κ are sentential variables, then the following are theorems:
    (i) $(\forall\sigma)\,(\forall\kappa)\,((T_r\text{'}(\sigma \wedge \kappa)\text{'} \Leftrightarrow (T_r\text{'}\sigma\text{'} \wedge T_r\text{'}\kappa\text{'}))$
    $(\forall\sigma)\,(\forall\kappa)\,((\neg T_r\text{'}(\sigma \wedge \kappa)\text{'} \Leftrightarrow ((\neg T_r\text{'}\sigma\text{'} \wedge T_r\text{'}\kappa\text{'}) \vee (T_r\text{'}\sigma\text{'} \wedge \neg T_r\text{'}\kappa\text{'}) \vee (\neg T_r\text{'}\sigma\text{'} \wedge \neg T_r\text{'}\kappa\text{'})))$

(ii) $(\forall\sigma)\ (\forall\kappa)\ ((T_r\text{'}(\sigma \lor \kappa)\text{'} \Leftrightarrow ((T_r\text{'}\sigma\text{'} \land T_r\text{'}\kappa\text{'}) \lor (\neg T_r\text{'}\sigma\text{'} \land T_r\text{'}\kappa\text{'}) \lor (T_r\text{'}\sigma\text{'} \land \neg T_r\text{'}\kappa\text{'})))$
$(\forall\sigma)\ \forall\kappa)\ ((\neg T_r\text{'}(\sigma \lor \kappa)\text{'} \Leftrightarrow (\neg T_r\text{'}\sigma\text{'} \land \neg T_r\text{'}\kappa\text{'}))$

(iii) $(\forall\sigma)\ (\forall\kappa)\ ((T_r\text{'}(\sigma \Leftrightarrow \kappa)\text{'} \Leftrightarrow ((T_r\text{'}\sigma\text{'} \land T_r\text{'}\kappa\text{'}) \lor (\neg T_r\text{'}\sigma\text{'} \land \neg T_r\text{'}\kappa\text{'})))$
$(\forall\sigma)\ \forall\kappa)\ ((\neg T_r\text{'}(\sigma \Leftrightarrow \kappa)\text{'} \Leftrightarrow ((\neg T_r\text{'}\sigma\text{'} \land T_r\text{'}\kappa\text{'}) \lor (T_r\text{'}\sigma\text{'} \land \neg T_r\text{'}\kappa\text{'})))$

(iv) $(\forall\sigma)\ (\forall\kappa)\ ((T_r\text{'}(\sigma \Rightarrow \kappa)\text{'} \Leftrightarrow (T_r\text{'}\sigma\text{'} \land T_r\text{'}\kappa\text{'}) \lor (\neg T_r\text{'}\sigma\text{'} \land T_r\text{'}\kappa\text{'}) \lor (\neg T_r\text{'}\sigma\text{'} \land \neg T_r\text{'}\kappa\text{'})))$
$(\forall\sigma)(\forall\kappa)\ ((\neg T_r\text{'}(\sigma \Rightarrow \kappa)\text{'} \Leftrightarrow (T_r\text{'}\sigma\text{'} \land \neg T_r\text{'}\kappa\text{'}))$

(v) $(\forall\sigma)\ (T_r\text{'}\sigma\text{'} \lor \neg T_r\text{'}\sigma\text{'})$

Theorem AP5 If σ and κ are sentential variables, then

$$(\forall\sigma)\ (T_r\text{'}\sigma\text{'} \Leftrightarrow (\forall\kappa)\ (\text{'}\sigma\text{'} = \text{'}\kappa\text{'} \Rightarrow \kappa)).$$

Observation: Theorem AP5 shows that in the context of the classical sentential calculus with substitutional quantifiers (and equality), the truth predicate is dispensable as primitive notation.

Theorem AP6 If σ and κ are sentential variables, and ω is any wff, then
(i) $(\forall\sigma)\omega \Leftrightarrow \neg(\exists\kappa)\neg\omega$, and
(ii) $(\exists\kappa)\omega \Leftrightarrow \neg(\forall\sigma)\neg\omega$.

We next turn to $\mathcal{L}_1^{**}$ to illustrate the small amount of proof theory and semantics available here. Notice that substitutional quantification gives us the capacity to characterize inductively sets of sentences without set theory; this is because the substitutional variables are fixed in their interpretation. First we shall build up to a definition of *tautology* (Definition AP16) by means of a couple of straightforward definitions and one axiom; and then we shall note briefly what is possible by way of proof theory.

Definition AP14 PS'σ' $\equiv_{df}$ $\neg(\exists\alpha)\ (\exists\beta)\ (\text{'}\sigma\text{'} = \text{'}\neg\alpha\text{'} \lor \text{'}\sigma\text{'} = \text{'}\alpha \Rightarrow \beta\text{'})$. ("'$\sigma$' is a primitive sentence.")

Definition AP15 The following axiom is called *the same structure axiom* (SSA), and it governs the predicate symbol P of $\mathcal{L}_1^{**}$ (which we have singled out as an interpreted predicate for the moment). In giving the axiom, we shall avail ourselves of the abbreviation made possible by Definition AP14.

(SSA) 'σ'S'δ' $\Leftrightarrow$ ((PS'σ' $\land$ PS'δ') $\lor$ $(\exists\alpha)\ (\exists\beta)\ (\text{'}\sigma\text{'} = \text{'}\neg\alpha\text{'} \land \text{'}\delta\text{'} = \text{'}\neg\beta\text{'} \land$ 'a'S'β') $\lor$ $(\exists\alpha)\ (\exists\beta)\ (\exists\gamma)\ (\exists\zeta)\ (\text{'}\sigma\text{'} = \text{'}\alpha \Rightarrow \beta\text{'} \land \text{'}\delta\text{'} = \text{'}\gamma \Rightarrow \zeta\text{'} \land$ 'a'S'γ' $\land$ 'β'S'ζ')).

Lastly, we give the definition of a tautology.

Definition AP16 TAUT'σ' $\equiv_{df}$ ($\forall\alpha$) ('α'S'σ' $\Rightarrow$ T$_r$'α').

Observation: Notice that Definition AP16 picks out every sentence if every primitive sentence is true.

Next we give the definition of an axiom. For illustrative purposes we choose not the particular proof procedure of L$_1$* itself, but another:

Definition AP17 Ax'σ' $\equiv_{df}$ ($\exists\alpha$) ($\exists\beta$) ('σ' = '($\alpha \Rightarrow (\beta \Rightarrow \alpha)$)') $\vee$ ($\exists\alpha$) ($\exists\beta$) ($\exists\gamma$) ('σ' = '((($\alpha \Rightarrow (\beta \Rightarrow \gamma)$) $\Rightarrow$ (($\alpha \Rightarrow \beta$) $\Rightarrow$ ($\alpha \Rightarrow \gamma$)))') $\vee$ ($\exists\alpha$) ($\exists\beta$) ('σ' = '(($\neg\alpha \Rightarrow \neg\beta$) $\Rightarrow$ ($\beta \Rightarrow \alpha$))'). ("'σ' is an axiom.")

Observations:

(i) We essentially have everything needed for the semantics of the classical sentential calculus. Theorem AP4 gives us the truth tables, and Definition AP17 gives us the notion of a tautology. Notice that

Theorem AP4 is far more significant than Theorem AP3, despite the fact that the latter is a version of the Tarskian definition for the connectives. The reason is that, as we will see, Theorem AP3 holds of the intuitionistic sentential calculus (turning as it does only on Theorem AP2), but Theorem AP4, with its detailed description of how the truth of sentences turns on the truth of their subsentential components, does not.

(ii) Proof theory is less successful. The problem is that we need extra resources for describing proofs, because proofs are open-ended in length. What can be defined (although we skip the details) is a hierarchy of proof predicates: "'σ' is a one-line proof," "'σ' is a two-line proof," and so on. But the definition of "proof," and thence "theorem," eludes us.

Model Theory

Because L* is wedded so closely to the underlying system L, it is not possible to give a general proof of the adequacy of (I1)–(I10); indeed, in rather pathological circumstances they are *not* adequate. Instead, therefore, we shall content ourselves with illustrating their adequacy in the classical case. To this end, we give a model theory for L$_1$*. The language we supply the model theory for is $\mathcal{L}_1$** plus Ξ-*wffs;* in doing this we automatically handle the cases of $\mathcal{L}_1$* and $\mathcal{L}_1$**.

Definition AP18 A model M for $\mathscr{L}_1$** is a mapping from the set of primitive sentences to $\{t, f\}$, the set of truth values; and, for each n, a mapping of the set of n-place predicate variables of $\mathscr{L}_1$** to the set of subsets of n-tuples of sentences of $\mathscr{L}_1$.

We next turn to the notion of *truth in a model*. Because our quantifiers are substitutional, we don't define the notion via satisfaction.

Definition AP19 Let M be a model for $\mathscr{L}_1$**.

(a) σ, a primitive sentence of $\mathscr{L}_1$, is true in M iff M maps σ to t.

(b) P'σ_1'...'σ_n', an n-place predicate followed by n terms, where $\sigma_1, \ldots, \sigma_n$ are sentences of $\mathscr{L}_1$, is true in M iff the subset of $\mathscr{L}_1$ that M maps P to contains the n-tuple $(\sigma_1, \ldots, \sigma_n)$.

(c) ('σ' = 'γ'), where σ and γ are sentences, is true in M iff σ is γ.

(d) $\neg\sigma$, where σ is a sentence, is true in M iff σ is not true in M.

(e) $(\sigma \Rightarrow \gamma)$, where σ and γ are sentences, is true in M iff either σ is not true in M or γ is true in M.

(f) T$_r$'σ', where σ is a sentence of $\mathscr{L}_1$, is true in M iff σ is true in M.

(g) $(\exists\alpha)$ σ, where α is a sentential variable, and σ is a wff with only α free in it, is true in M iff there is a sentence γ of $\mathscr{L}_1$, such that the sentence gotten by substituting γ for all free occurrences of α in σ is true in M.

(h) $(\forall\alpha)\sigma$, where α is a sentential variable, and σ is a wff with only α free in it, is true in M iff every sentence gotten by substituting any sentence γ of $\mathscr{L}_1$, for all free occurrences of α in σ is true in M.

(i) $\varsigma(\Xi)$, where ς is a wff with only the sentential variable α free in it, is true in M iff any sentence gotten by substituting any sentence of $\mathscr{L}_1$ for α in ς is true in M.

(j) $(\varsigma(\Xi)/\rho_1, \ldots, \rho_n)$, where ς is a wff with only the sentential variable α free in it, is true in M iff any sentence gotten by substituting a sentence of $\mathscr{L}_1$ (except for $\rho_1, \ldots, \rho_n$) for α in ς is true in M.

Completeness and Consistency

Theorem AP7 (I1)–(I10), (I$_1$1), and (I$_1$2) are the truth preserving with respect to the notions of *model* and *truth in a model* defined via Definitions AP18 and AP19.

Theorem AP8 Let a set of sentences Ω be consistent with respect to (I1)–(I10), (I$_1$1), and (I$_1$2) (where (I10) can be used infinitely often). Then there is a model M with respect to which all the sentences of Ω are true.

Proof Sketch We give what is essentially a Henkin proof. Using Theorems AP5 and AP6, it suffices to treat $\mathcal{L}_1$** as if it lacks T_r and the universal substitutional quantifier. Because Ω can be infinite, we will assume that every Ξ-wff has been replaced by the set of sentences it represents, we shall take (I1)–(I10) to be written in our primitive notation, and in particular, we will treat (I10) as an infinite derivation rule.

The first step, therefore, is to expand Ω into a maximally consistent set of sentences. To this end, consider a list of all the sentences of $\mathcal{L}_1$**: $\omega_1, \dots, \omega_n \dots$. We define the following sequence of sets of sentences:

$\Omega_1 = \Omega$,

(i) If ω_n is not a sentence of the form $(\exists\alpha)\sigma$, then $\Omega_{n+1} = \Omega_n$ plus ω_n, if the resulting set of sentences is syntactically consistent.
$\Omega_{n+1} = \Omega_n$ plus $\neg\omega_n$, otherwise.

(ii) If ω_n is a sentence of the form $(\exists\alpha)\sigma$, then $\Omega_{n+1} = \Omega_n$ plus ω_n and one sentence of the form $\sigma[\omega]$ where $\sigma[\omega]$ is gotten by substituting some sentence ω of $\mathcal{L}_1$ for α in σ, if the resulting set of sentences is syntactically consistent,
$\Omega_{n+1} = \Omega_n$ plus $\neg\omega_n$, and *every* sentence of the form $\neg\sigma[\omega]$, where ω is a sentence of $\mathcal{L}_1$, otherwise.

Let $\Omega^{\#} = \mathsf{U}_n\Omega_n$. Finally, add every Ξ-*wff* which is consistent with $\Omega^{\#}$. Call the resulting set of sentences Ω^*.

Claim: Ω^* is a maximally syntactically consistent set of sentences.

Proof Sketch: Where the procedure applied to an Ω_i is (i), this is shown in the standard way. Use (I10) to show that the procedure outlined in (ii) cannot get us in trouble. To show that taking the union of the Ω_n doesn't cause trouble either, note that the procedure outlines in (ii) covers every potentially troublesome case that might arise over (I10).

Now we use Ω^* to define the model M. A primitive sentence ω is mapped to t iff it is in Ω^*. An n-tuple of sentences $(\sigma_1, \dots, \sigma_n)$ is in the set of n-tuples an n-place predicate P is mapped to iff the sentence $P'\sigma_1'\dots'\sigma_n'$ is in Ω^*.

Claim: A sentence ω is true in M iff it is in Ω^*.

Proof Sketch: We show this by induction using Definition AP19 and (I1)–(I10).

(a) and (b) hold by definition. (c), (d), (e), (i), and (j) are obvious.

(g) $(\exists\alpha)\sigma$ is true iff there is a sentence γ of $\mathcal{L}_1$, such that the sentence gotten by substituting γ for all free occurrences of α is σ is true but (by assumption) this hold iff γ is in Ω^* and (by construction of Ω^*) iff $(\exists\alpha)$ σ is in Ω^*.

This completes the proof.

The Standard Predicate Calculus

The case of the standard predicate calculus will be covered in a more cursory fashion because many of the moves are identical to the case of the classical sentential calculus.

We now take $\mathcal{L}$ to be the language of the classical predicate calculus and L to be the classical logical truths (we call them $\mathcal{L}_2$ and L_2).

The vocabulary of $\mathcal{L}_2$ is the following: n-place predicates: **P, Q, R, . . .** , an existential quantifier: **∃**, variables: **x, y, z, . . .** , individual constants: **a, b, c, . . .** , parentheses: **(,)**, connectives: **⇒, ¬**.

Observation: We distinguish the vocabulary of $\mathcal{L}_2$ from $\mathcal{L}_2$* ($\mathcal{L}_2$**) where necessary by putting the terminology of $\mathcal{L}_2$ in boldface.

Definition AP20 We adopt the standard formation rules for the classical predicate calculus. "∧," "∨," "⇔," and "∀" are introduced via definitions in the usual way. Note that there is a distinction in $\mathcal{L}_2$ between sentences and wffs.

Definition AP21 As inference rules for L_2, we have $(I_1 1)$, $(I_1 2)$ (which we call hereafter $(I_2 1)$ and $(I_2 2)$), and:

$(I_2 3)$ If σ is any variable, ς any wff, and ς[κ] a wff differing from ς, in that every free occurrence of σ has been replaced by κ, which is either an individual constant or a variable, and further if ς has the same number of free variables as ς[κ] if κ is a variable, we have

$$ς[κ] \vdash (∃σ)ς.$$

$(I_2 4)$ If σ is any variable, ς any wff, and ς[κ] a wff differing from ς in that every free occurrence of σ has been replaced by κ, an individual constant or a variable, and further if ς has the same number of free variables as ς[κ] if κ is a variable, and σ does not occur free in any of the sentences whose line numbers appear in the premise list of $(∃σ)ς$, we have

$$(∃σ)ς \vdash ς[κ].$$

Definition AP22 An L2-derivation is a finite $\mathcal{L}_2$*-derivation in which every wff of the derivation is a sentence of $\mathcal{L}_2$.

Because L_2 strictly contains L_1, Theorems AP1–AP5 all hold here too. Furthermore, we can define as before the set of tautologies, which is a strict subclass of the logical truths. But this is as far as we can take the semantics. To go further, it is necessary to be able to refer substitutionally to wffs that are not sentences, such as "Px," and to be able to talk not just about truth but about satisfaction as well. We skip further details.

Notice that the model theory for $\mathcal{L}_1{}^{**}$ essentially amounted to models for the sentential calculus that were augmented with mappings from the n-place predicate variables of $\mathcal{L}_1{}^{**}$ to the set of subsets of n-tuples of sentences of $\mathcal{L}_1$. This clearly generalizes to the standard predicate calculus.

Definition AP23 A model M for $\mathcal{L}_2{}^{**}$ is a model for M^T for $\mathcal{L}_2$ plus, for each n, a mapping of the set of n-place predicate variables of $\mathcal{L}_2{}^{**}$ to the set of subsets of n-tuples of sentences of $\mathcal{L}_2$.

We now go on to give the the notion of *truth in a model*. In doing so we presuppose the notion of *truth in a model of $\mathcal{L}_2$*, that is, the notion of truth for the standard predicate calculus.

Definition AP24 Let M be a model for $\mathcal{L}_2{}^{**}$.
(a) σ, a sentence of $\mathcal{L}_2$, is true in M iff it is true in M^T.
(b)–(j) are as in Definition AP19.

Theorem AP9 (I1)–(I10), $(I_2 1)$–$(I_2 4)$ are truth preserving with respect to the notions of *model* and *truth in a model* defined via Definitions AP23 and AP24.

Theorem AP10 Let a set of sentences Ω be consistent with respect to (I1)–(I10), $(I_1 1)$–$(I_2 4)$ (Where (I10) can be used infinitely often). Then there is a model M with respect to which all the sentences of Ω are true.

Proof Sketch We use a (slightly modified) Henkin proof. First expand Ω by adding new individual constants to $\mathcal{L}_2$, and adding sentences of the form $(\exists σ)ς \Rightarrow ς[κ]$, where κ is a new constant substituted for free σ in ς, for each wff of the expanded language. Then expand the result to a maximally syntactically consistent set of sentences as was done for Theorem AP8.

The Intuitionistic Sentential Calculus

Our last example is the intuitionistic sentential calculus. The vocabulary of $\mathcal{L}_3$ is the following: syntactically primitive sentences: $S_1, \ldots$,' $S_n, \ldots$, parentheses: (,), a propositional constant: $\bot$, and connectives: $\Rightarrow, \wedge, \vee$.

Observations:

(i) As with the classical sentential calculus, we use syntactically primitive sentences rather than sentential variables.

(ii) We introduce "¬," and "⇔," the usual way.

Definition AP25 The following are the formation rules for $\mathcal{L}_3$:

($\mathcal{L}_3$1) If ς is a primitive sentence, then it is a wff.

($\mathcal{L}_3$2) $\bot$ is a wff.

($\mathcal{L}_3$3) If ς and ρ are wffs, then $(\varsigma \Rightarrow \rho)$, $(\varsigma \wedge \rho)$, and $(\varsigma \vee \rho)$ are wffs.

($\mathcal{L}_3$4) Something is a wff if and only if it is so by virtue of clauses ($\mathcal{L}_3$1)–($\mathcal{L}_3$3).

Definition AP26 The following rules (adapted from Van Dalen 1986) are called the inference rules of L_3. We use "antecedent," "consequent," and "clause," as in AP4.

($I_3$1) If ς and ρ are wffs, then

$$\varsigma, \rho \vdash (\varsigma \wedge \rho),$$
$$(\varsigma \wedge \rho) \vdash \varsigma,$$

and

$$(\varsigma \wedge \rho) \vdash \rho.$$

($I_3$2) If ς and ρ are wffs, then

$$\varsigma \vdash (\varsigma \vee \rho)$$

and

$$\rho \vdash (\varsigma \vee \rho).$$

($I_3$3) If ς, ρ, and σ are wffs, then we have

$$(\varsigma \Rightarrow \sigma), (\rho \Rightarrow \sigma) \vdash ((\varsigma \vee \rho) \Rightarrow \sigma).$$

($I_3$4) If ς and ρ are wffs, then

$$\varsigma, (\varsigma \Rightarrow \rho) \vdash \rho.$$

($I_3$5) If ς and ρ are wffs, then

$$\rho \vdash (\varsigma \Rightarrow \rho).$$

As before, this is a premise-deleting rule. The premise numbers of the new line are all of those of the previous line, except (if desired) any that is the line number of a line on which ς appears.

($I_3$6) If ς is a wff, then

$$\bot \vdash \varsigma.$$

We next describe a few of the definitions and theorems that are available in L_3* and mention a few contrasts with L_1*.

First off, a theorem corresponding to the classical Theorem AP1 is not available here. We do not have intuitionistic theorems corresponding to Theorems AP2 and AP3, however. Nothing corresponding to Theorem AP4, which describes truth tables, is available. The best we can do are theorems gotten from those of the intuitionistic sentential calculus by substitution of $T_r'\sigma'$ for σ – for example: (∀σ) (∀κ) ((¬T_r‘(σ ∧ κ)’ ⇔ (T_r‘σ’ ⇒ ¬T_r‘κ’)). The correspondent to Theorem AP5 is available, but because the underlying connectives are intuitionistic, the correspondent to Theorem AP6 is not. Only those intuitionistic relationships among the quantifiers such as (∀σ)ω ⇔ ¬(∃κ)ω can be derived.

Concluding Remarks

(i) Theorem AP2, and those results depending on it by the substitution of equivalents for equivalents are fairly robust; that is, they tend to be present wherever something possessing the properties of "⇔" is available. This does seem to show that the content of Theorem AP3 is slight at best.

(ii) Usually, the semantics and proof theory in a metalanguage are built on concatenation theory. Concatenation theory, however, is *quite* powerful – and we have seen reasons to suspect that perhaps the substitutional idiom will enable us to do the job with less. This is not the place to pursue that tantalizing possibility in detail, but indeed, with only a modest increase in the referential powers of the language the truth predicate is couched in, it is possible to go much further than we have gone.

(iii) Recall my aim in writing this appendix. I wanted to show the existence of a truth predicate minimal in its philosophical and technical presuppositions; I wanted this because such a truth predicate is susceptible to use in a variety of contexts, classical and otherwise. But I do not assume that substitutional quantification comes without "ontological" costs, although I have not tried to evaluate them.

Bibliography

Aczel, Peter. 1987. *Lectures on nonwellfounded sets.* CSLI Lecture Notes No. 9. Stanford, CA: CSLI/Stanford University.

Alston, William P. 1980. Level confusions in epistemology. *Midwest Studies in Philosophy, 5,* 135–50. Reprinted in William P. Alston, *Epistemic justification* (pp. 153–71). Ithaca, NY: Cornell University Press, 1989.

Atiyah, Michael. 1990. *The geometry and physics of knots.* Cambridge: Cambridge University Press.

Azzouni, Jody. 1990. Truth and convention. *Pacific Philosophical Quarterly, 71,* 81–102.

1991. A simple axiomatizable theory of truth. *Notre Dame Journal of Formal Logic, 32,* no. 3, 458–93.

1992. A priori truth. *Erkenntnis, 37,* 327–46.

Barwise, J., and Robin Cooper. 1981. Generalized quantifiers and natural language. *Linguistics and Philosophy, 4,* 159–219.

Barwise, J., and J. Etchemendy. 1987. *The liar: An essay on truth and circularity.* New York: Oxford University Press.

Barwise, J., and S. Feferman (Eds.). 1985. *Model-theoretic logics.* Berlin: Springer-Verlag.

Bay, Mel. 1990. *Mel Bay's modern guitar method, Grade 1.* Pacific, MO: Mel Bay Publications.

Belnap, Nuel D., and Dorothy Grover. 1973. Quantifying in and out of quotes. In Hughes Leblanc (Ed.), *Truth, syntax, and modality* (pp. 17–47). Dordrecht: Reidel. Reprinted in Dorothy Grover, *A prosentential theory of truth* (pp. 244–75). Princeton, NJ: Princeton University Press, 1992.

Benacerraf, Paul. 1965. What numbers could not be. *Philosophical Review, 74,* 47–73. Reprinted in Paul Benacerraf and Hilary Putnam (Eds.), *Philosophy of mathematics,* 2nd ed. (pp. 272–94). Cambridge: Cambridge University Press, 1983.

1973. Mathematical truth. *Journal of Philosophy, 70,* 661–80. Reprinted in Paul Benacerraf and Hilary Putnam (Eds.), *Philosophy of mathematics,* 2nd ed. (pp. 403–20). Cambridge: Cambridge University Press, 1983.

1985. Skolem and the skeptic. *Proceedings of the Aristotelian Society,* Suppl. 59, 85–115.

Benacerraf, Paul, and Hilary Putnam (Eds.). 1983. Introduction. In *Philosophy of Mathematics*, 2nd ed. (pp. 1–37). Cambridge: Cambridge University Press.

Bonevac, D. A. 1982. *Reduction in the abstract sciences.* Indianapolis, IN: Hackett.

Boolos, George S. 1975. On second-order logic. *Journal of Philosophy, 72,* 509–27.

1984. To be is to be a value of a variable (or to be some values of some variables). *Journal of Philosophy, 81,* 430–49.

Boolos, George S., and Richard C. Jeffrey. 1989. *Computability and logic,* 3rd ed. Cambridge: Cambridge University Press.

Bos, H. J. M. 1984. Arguments on motivation in the rise and decline of a mathematical theory; the 'construction of equations,' 1637–ca. 1750. *Archive for History of Exact Sciences, 30,* 331–80.

Burgess, John P. 1983. Why I am not a nominalist. *Notre Dame Journal of Formal Logic, 24,* 93–105.

Carnap, Rudolf. 1949. Truth and confirmation. In Herbert Feigl and Wilfred Sellars (Eds.), *Readings in philosophical analysis* (pp. 119–27). New York: Appleton-Century-Crofts.

1956. Empiricism, semantics, and ontology. In *Meaning and necessity* (pp. 205–21). Chicago: University of Chicago Press.

Chang, C. C., and Jerome H. Keisler. 1973. *Model theory.* Amsterdam: North-Holland.

Chihara, Charles S. 1990. *Constructibility and mathematical existence.* Oxford: Oxford University Press.

Church, Alonzo. 1956. *Introduction to mathematical logic,* Vol. 1. Princeton, NJ: Princeton University Press.

Churchland, Paul M. 1979. *Scientific realism and the plasticity of mind.* Cambridge: Cambridge University Press.

Cohn, Harvey. 1980. *Advanced number theory.* New York: Dover. Originally published as *A second course in number theory.* New York: Wiley, 1962.

Condon, Edward U., and Philip M. Morse. 1929. *Quantum mechanics.* New York: McGraw-Hill.

Corcoran, John. 1980. Categoricity. *History and Philosophy of Logic, 1,* 187–207.

1981. From categoricity to completeness. *History and Philosophy of Logic, 2,* 113–19.

Crowe, Michael J. 1988. Ten misconceptions about mathematics and its history. In William Aspray and Philip Kitcher (Eds.), *History and philosophy of modern mathematics* (pp. 260–77). Minnesota Studies in the Philosophy of Science 9. Minneapolis: University of Minnesota Press.

Daston, Lorraine. 1988. Fitting numbers to the world: The case of probability theory. In William Aspray and Philip Kitcher (Eds.), *History and philosophy of modern mathematics* (pp. 221–37). Minnesota Studies in the Philosophy of Science 9. Minneapolis: University of Minnesota Press.

Davidson, Donald. 1986a. Reality without reference. In *Inquiries into truth & interpretation* (pp. 215–25). Oxford: Clarendon.

1986b. The inscrutability of reference. In *Inquiries into truth & interpretation* (pp. 227–41). Oxford: Clarendon.

1986c. On the very idea of a conceptual scheme. In *Inquiries into truth &* *interpretation* (pp. 183–98). Oxford: Clarendon.

Davis, Martin. 1977. *Applied nonstandard analysis.* New York: Wiley.

Descartes, René. 1931a. Rules for the direction of the mind. In *The philosophical works of Descartes,* trans. by Elizabeth S. Haldane and G. R. T. Ross (pp. 1–77). London: Cambridge University Press.

1931b. Meditations on first philosophy. In *The philosophical works of Descartes,* trans. by Elizabeth S. Haldane and G. R. T. Ross (pp. 131–99). London: Cambridge University Press.

Detlefsen, Michael. 1979. On interpreting Gödel's second theorem. *Journal of Philosophical Logic, 8,* 297–313.

Detlefsen, Michael, and Mark Luker. 1980. The four-color theorem and mathematical proof. *Journal of Philosophy, 77,* 803–20.

Devitt, Michael. 1981. *Designation.* New York: Columbia University Press.

1983. Realism and the renegade Putnam. *Noûs, 17,* 291–301.

Donnellan, Keith S. 1966. Reference and definite descriptions. *Philosophical Review, 75,* 281–304. Reprinted in Stephen P. Schwartz (Ed.), *Naming, necessity, and natural kinds* (pp. 42–65). Ithaca, NY: Cornell University Press.

Dretske, Fred I. 1981. *Knowledge and the flow of information.* Cambridge, MA: MIT Press.

Dummett, Michael. 1963. Realism. In *Truth and other enigmas* (pp. 145–65). Cambridge, MA: Harvard University Press.

1977. *Elements of intuitionism.* Oxford: Oxford University Press.

Dunn, Michael J., and Nuel D. Belnap, Jr. 1968. The substitution interpretation of the quantifiers. *Noûs, 2,* 177–85.

Enderton, Herbert B. 1972. *A mathematical introduction to logic.* New York: Academic Press.

Feferman, Solomon. 1960. Arithmetization of meta-mathematics in a general setting. *Fundamenta Mathematicae, 64.*

1982. Toward useful type-free theories. I. In Robert L. Martin (Ed.), *Recent essays on truth and the liar paradox* (pp. 237–87). Oxford: Oxford University Press, 1984.

Field, Hartry. 1972. Tarski's theory of truth. *Journal of Philosophy, 64,* 347–75.

1975. Conventionalism and instrumentalism in semantics. *Noûs, 9,* 376–406.

1980. *Science without numbers.* Princeton, NJ: Princeton University Press.

1989. *Realism, mathematics & modality.* Oxford: Blackwell.

Fodor, Jerry A. 1987. *Psychosemantics.* Cambridge, MA: MIT Press.

Frege, G. 1894. Review of Dr. E. Husserl's *Philosophy of arithmetic. Mind, 81,* 321–37.

1953. *Foundations of arithmetic,* trans. J. L. Austin. Oxford: Basil Blackwell.

1967. *The basic laws of arithmetic,* ed. and trans. Montgomery Furth. Berkeley: University of California Press.

Friedman, Michael. 1988. Logical truth and analyticity in Carnap's 'logical syntax of language.' In William Aspray and Philip Kitcher (Eds.), *History and philosophy of modern mathematics* (pp. 82–94). Minnesota Studies in the Philosophy of Science 9. Minneapolis: University of Minnesota Press.

238   BIBLIOGRAPHY

Glymour, Clark. 1982. Conceptual scheming, or confessions of a metaphysical realist. *Synthese, 51,* 169–80.

Gödel, K. 1947. What is Cantor's continuum problem?" Revised version reprinted in Paul Benacerraf and Hilary Putnam (Eds.), *Philosophy of mathematics,* 2nd ed. (pp. 470–85). Cambridge: Cambridge University Press, 1983.

Goldman, Alvin. 1986. *Epistemology and cognition.* Cambridge, MA: Harvard University Press.

Gottlieb, Dale. 1980. *Ontological economy.* Oxford: Oxford University Press.

Grattan-Guinness, I. 1970. *The development of the foundations of analysis from Euler to Riemann.* Cambridge, MA: MIT Press.

    1973. Review of Kline's *Mathematical thought from ancient to modern times. Science, 180,* May 11, 627.

Grice, Paul, and P. F. Strawson. 1956. In defense of a dogma. In Paul Grice, *Studies in the way of words* (pp. 196–212). Cambridge, MA: Harvard University Press, 1989.

Grover, Dorothy. 1990. Two deflationary truth theories. In Dorothy Grover, *A prosentential theory of truth* (pp. 215–33). Princeton, NJ: Princeton University Press, 1992.

Hacking, Ian. 1983. *Representing and intervening.* Cambridge: Cambridge University Press.

Hardy, G. H. 1967. *A mathematician's apology.* Cambridge: Cambridge University Press (1st ed. 1940).

Hellman, Geoffrey. 1989. *Mathematics without numbers.* Oxford: Oxford University Press.

Hempel, Carl. 1965. Studies in the logic of confirmation. In *Aspects of scientific explanation* (pp. 3–51). New York: Free Press.

Henkin, Leon. 1950. Completeness in the theory of types. *Journal of Symbolic Logic, 15,* 81–91.

Herzberger, Hans G. 1980–1. New paradoxes for old. *Proceedings of the Aristotelian Society, NS 81,* 109–23.

    1982. Notes on naive semantics. *Journal of Philosophical Logic, 11,* 61–102. Reprinted in Robert L. Martin (Ed.), *Recent essays on truth and the liar paradox* (pp. 133–74). Oxford: Oxford University Press, 1984.

Hewitt, Edwin, and Karl Stromberg. 1965. *Real and abstract analysis.* New York: Springer-Verlag.

Hilbert, David. 1925. On the infinite. Reprinted in Paul Benacerraf and Hilary Putnam (Eds.), *Philosophy of mathematics,* 2nd ed. (pp. 183–201). Cambridge: Cambridge University Press, 1983.

Hille, Einar. 1973. *Analytic function theory,* 2nd ed., 2 vols. New York: Chelsea.

Hodes, Harold T. 1984. Logicism and the ontological commitments of arithmetic. *Journal of Philosophy, 81,* 123–49.

Husserl, Edmund. 1970. *Logical investigations,* Vol. I. New York: Humanities Press.

Isles, David. A finite analog to the Löwenheim–Skolem theorem. Unpublished.

    1992. What evidence is there that 2^{65536} is a natural number? *Notre Dame Journal of Formal Logic, 33,* no. 4 (Fall), 465–80.

Kant, Immanuel. 1965. *Critique of pure reason*, trans. by Norman Kemp Smith. New York: St. Martin's.

Katz, Jerrold J. 1979. Semantics and conceptual change. *Philosophical Review, 88*, 327–65.

1981. *Language & other abstract objects*, Totowa, NJ: Rowman and Littlefield.

Kaye, Richard. 1991. *Models of Peano arithmetic*. Oxford: Oxford University Press.

Kitcher, Philip. 1978. The plight of the platonist. *Noûs, 12*, 119–36.

1984. *The nature of mathematical knowledge*. Oxford: Oxford University Press.

Kleene, Stephen Cole, and Richard Eugene Vesley. 1965. *The foundations of intuitionistic mathematics*. Amsterdam: North-Holland.

Kline, Morris. 1972. *Mathematical thought from ancient to modern times*. New York: Oxford University Press.

1980. *Mathematics: The loss of certainty*. Oxford: Oxford University Press.

Kripke, Saul. 1975. Outline of a theory of truth. *Journal of Philosophy, 72*, 690–716. Reprinted in Robert L. Martin (Ed.), *Recent essays on truth and the liar paradox* (pp. 53–81). New York: Oxford University Press, 1984.

1980. *Naming and necessity*. Cambridge, MA: Harvard University Press.

1982. *Wittgenstein on rules and private language*. Cambridge, MA: Harvard University Press.

Lakatos, Imre. 1976a. *Proofs and refutations: The logic of mathematical discovery*, eds. John Worrall and Elie Zahar. Cambridge: Cambridge University Press.

1976b. Falsification and the methodology of scientific research programmes. In Imre Lakatos and Alan Musgrave (Eds.), *Criticism and the growth of knowledge*. Cambridge: Cambridge University Press.

Langer, R. E. 1947. Fourier's series: The genesis and evolution of a theory. Supplement to *American Mathematical Monthly, 54*, no. 7.

Leblanc, Hughes. 1983. Alternatives to standard first-order semantics. In D. Gabbay and F. Guenthner (Eds.), *Handbook of philosophical logic* (Vol. I, pp. 189–274). Boston: D. Reidel.

Leeds, Stephen. 1978. Theories of reference and truth. *Erkenntnis, 13*, 111–29.

Lewis, David. 1984. Putnam's paradox. *Australasian Journal of Philosophy, 62*, 221–36.

Luce, Lila. 1989. Platonism from an empiricist point of view. *Philosophical Topics, 17*, 109–28.

Maddy, Penelope. 1980. Perception and mathematical intuition. *Philosophical Review, 89*, 163–96.

1989. The roots of contemporary platonism. *Journal of Symbolic Logic, 54*, 1121–44.

1990. *Realism in mathematics*. Oxford: Oxford University Press.

1991. Philosophy of mathematics: Prospects for the 1990s. *Synthese, 88*, 155–64.

Malament, D. 1982. Review of Field's "Science without numbers." *Journal of Philosophy, 79*, 523–34.

Martin, Robert L. 1984. *Recent essays on truth and the liar paradox*. New York: Oxford University Press.

Mates, Benson. 1965. *Elementary logic.* Oxford: Oxford University Press.

Mendelson, Elliott. 1990. Second thoughts about Church's thesis and mathematical proofs. *Journal of Philosophy, 87,* 225–33.

Moore, Gregory H. 1980. Beyond first-order logic: The historical interplay between mathematical logic and axiomatic set theory. *History and Philosophy of Logic, 1,* 95–137.

  1982. *Zermelo's axiom of choice: Its origins, development, and influence.* New York: Springer-Verlag.

Myhill, John. 1951. On the ontological significance of the Löwenheim–Skolem theorem. In M. White (Ed.), *Academic freedom, logic and religion* (pp. 57–70). Reprinted in I. Copi and J. Gould (Eds.), *Contemporary readings in logical theory* (pp. 40–54). New York: Macmillan, 1967.

Nagel, Ernest. 1935. "Impossible numbers": A chapter in the history of modern logic. *Studies in the history of ideas, 3,* 427–74. Reprinted in Ernest Nagel, *Teleology revisited* (pp. 166–94). New York: Columbia University Press, 1979.

Nelson, Edward. Taking formalism seriously. *Proceedings in Logic, Method and Philosophy of Science* (Uppsala), forthcoming.

Nevanlinna, Rolf, and V. Paatero. 1969. *Introduction to complex analysis.* Reading, MA: Addison-Wesley.

Newman, James. (Ed.). 1956. Srinivasa Ramanujan. In James Newman (Ed.), *The world of mathematics* (Vol. 1, pp. 368–76).

Parikh, Rohit. 1971. Existence and feasibility in arithmetic. *Journal of Symbolic Logic, 36,* 494–508.

Parsons, Charles. 1974a. Informal axiomatization, formalization, and the concept of truth. In Charles Parsons, *Mathematics in philosophy* (pp. 71–91). Ithaca, NY: Cornell University Press, 1983.

  1974b. Sets and classes. In Charles Parsons, *Mathematics in philosophy* (pp. 209–20). Ithaca, NY: Cornell University Press, 1983.

  1979–80. Mathematical intuition. *Proceedings of the Aristotelian Society, NS 80,* 145–68. Reprinted in Palle Yourgrau (Ed.), *Demonstratives* (pp. 195–214). Oxford: Oxford University Press, 1990.

  1990. The structuralist view of mathematical objects. *Synthese, 84,* 303–46.

Pearce, David, and Veikko Rantala. 1982. Realism and formal semantics. *Synthese, 52,* 39–53.

Popper, Karl. 1965. *Conjectures and refutations: The growth of scientific knowledge.* New York: Harper Torchbooks.

Priest, Graham. 1979. Logic of paradox. *Journal of Philosophical Logic, 8,* 219–41.

Putnam, Hilary. 1962. The analytic and the synthetic. In Hilary Putnam, *Mind, language and reality: Philosophical papers* (Vol. 2, pp. 33–69). Cambridge: Cambridge University Press.

  1967. Mathematics without foundations. In Hilary Putnam, *Mathematics, matter and method: Philosophical papers* (Vol. 1, pp. 43–59). Cambridge: Cambridge University Press.

  1975a. What is mathematical truth? In Hilary Putnam, *Mathematics, matter and method: Philosophical papers* (Vol. 1, pp. 60–78). Cambridge: Cambridge University Press.

1975b. Is semantics possible. In Hilary Putnam, *Mind, language and reality: Philosophical papers* (Vol. 2, pp. 139–52). Cambridge: Cambridge University Press.

1975c. The meaning of "meaning." In Hilary Putnam, *Mind, language and reality: Philosophical papers* (Vol. 2, pp. 215–71). Cambridge: Cambridge University Press.

1975d. Philosophy and our mental life. In Hilary Putnam, *Mind, language and reality: Philosophical papers* (Vol. 2, pp. 291–303). Cambridge: Cambridge University Press.

1977. Models and reality. In Hilary Putnam, *Realism and reason* (pp. 1–25). Cambridge: Cambridge University Press.

1978a. Realism and reason. In Hilary Putnam, *Meaning and the moral sciences* (pp. 123–40). London: Routledge & Kegan Paul.

1978b. *Meaning and the moral sciences.* London: Routledge & Kegan Paul.

1979. Philosophy of mathematics: A report. In P. D. Asquith and R. N. Giere (Eds.), *Current research in philosophy of science* (pp. 386–98). East Lansing, MI: Philosophy of Science Association.

1981. *Reason, truth and history.* Cambridge: Cambridge University Press.

1983a. Why there isn't a ready-made world. In *Realism and reason* (pp. 205–28). Cambridge: Cambridge University Press.

1983b. Beyond historicism. In *Realism and reason* (pp. 287–303). Cambridge University Press.

1983c. Introduction: An overview of the problem. In *Realism and reason* (pp. vii–xviii). Cambridge: Cambridge University Press.

1983d. Two dogmas revisited. In *Realism and reason* (pp. 87–97). Cambridge: Cambridge University Press.

1983e. There is at least one a priori truth. In *Realism and reason* (pp. 98–114). Cambridge: Cambridge University Press.

1983f. Analyticity and apriority: Beyond Wittgenstein and Quine. In *Realism and reason* (pp. 115–38). Cambridge: Cambridge University Press.

1983g. Why reason can't be naturalized. In *Realism and reason* (pp. 229–47). Cambridge: Cambridge University Press.

1984. Is the causal structure of the physical itself something physical? In Peter A. French, Theodore E. Uehling, Jr., and Howard K. Wettstein (Eds.), *Midwest studies in philosophy* (Vol. 9, pp. 3–16). Minneapolis: University of Minnesota Press.

1986. Information and the mental. In Ernest Lepore (Ed.), *Truth and interpretation: Perspectives on the philosophy of Donald Davidson* (pp. 262–71). Oxford: Blackwell.

1987. *The many faces of realism.* La Salle, IL: Open Court.

1989. Model theory and the "factuality" of semantics. In Alexander George (Ed.), *Reflections on Chomsky* (pp. 213–32). Oxford: Blackwell.

Quine, W. V. O. 1936. Truth by convention. In W. V. O. Quine, *The ways of paradox and other essays,* rev. and enl. ed. (pp. 77–106). Cambridge, MA: Harvard University Press, 1976.

1940. *Mathematical logic.* New York: Norton.

1951a. Two dogmas of empiricism. In W. V. O. Quine, *From a logical point of view* (pp. 20–46). Cambridge, MA: Harvard University Press, 1980.

1951b. On what there is. In W. V. O. Quine, *From a logical point of view* (pp. 1–19). Cambridge, MA: Harvard University Press, 1980.

1954. Carnap and logical truth. In W. V. O. Quine, *The ways of paradox and other essays*, rev. and enl. ed. (pp. 107–32). Cambridge, MA: Harvard University Press, 1976.

1955. Posits and reality. In W. V. O. Quine, *The ways of paradox and other essays*, rev. and enl. ed. (pp. 246–54). Cambridge, MA: Harvard University Press, 1976.

1960. *Word and object.* Cambridge, MA: MIT Press.

1969a. *Set theory and its logic*, rev. ed. Cambridge, MA: Harvard University Press.

1969b. Ontological relativity. In W. V. O. Quine, *Ontological relativity & other essays* (pp. 26–68). New York: Columbia University Press.

1969c. Epistemology naturalized. In W. V. O. Quine, *Ontological relativity & other essays* (pp. 69–90). New York: Columbia University Press.

1969d. Natural kinds. In W. V. O. Quine, *Ontological relativity & other essays* (pp. 114–38). New York: Columbia University Press.

1970. *Philosophy of logic.* Englewood Cliffs, NJ: Prentice Hall.

1972. Vagaries of definition. In W. V. O. Quine, *The ways of paradox and other essays*, rev. and enl. ed. (pp. 246–54). Cambridge, MA: Harvard University Press, 1976.

1976. On multiplying entities. In W. V. O. Quine, *The ways of paradox and other essays*, rev. and enl. ed. (pp. 259–64). Cambridge, MA: Harvard University Press.

1981a. Empirical content. In W. V. O. Quine, *Theories and things* (pp. 24–30). Cambridge, MA: Harvard University Press.

1981b. What price bivalence? In W. V. O. Quine, *Theories and things* (pp. 31–37). Cambridge, MA: Harvard University Press.

1984. Review of Parsons's *Mathematics in philosophy. Journal of Philosophy, 81,* 783–94.

1986a. Reply to Harold N. Lee. In Lewis Edwin Hahn and Paul Arthur Schilpp (Eds.), *The philosophy of W. V. Quine* (pp. 315–18). La Salle, IL: Open Court.

1986b. Reply to Paul A. Roth. In Lewis Edwin Hahn and Paul Arthur Schilpp (Eds.), *The Philosophy of W. V. Quine* (pp. 459–61). La Salle, IL: Open Court.

1986c. Reply to Henryk Skolimowski. In Lewis Edwin Hahn and Paul Arthur Schilpp (Eds.), *The philosophy of W. V. Quine* (pp. 492–93). La Salle, IL: Open Court.

1990. *Pursuit of truth.* Cambridge, MA: Harvard University Press.

Reed, Michael, and Barry Simon. 1972. *Methods of modern mathematical physics,* Vol. 1. *Functional analysis.* New York: Academic Press.

Resnik, Michael D. 1974. On the philosophical significance of consistency proofs. *Journal of Philosophical Logic, 3,* 133–47.

1980. *Frege and the philosophy of mathematics.* Ithaca, NY: Cornell University Press.

1981. Mathematics as a science of patterns: Ontology and reference. *Noûs, 15,* 529–50.

1982. Mathematics as a science of patterns: Epistemology. *Noûs, 16,* 95–106.

1985. How nominalist is Hartry Field's nominalism? *Philosophical Studies, 47,* 163–81.

1988. Second-order logic still wild. *Journal of Philosophy, 85,* 75–87.

1989. Computation and mathematical empiricism. *Philosophical Topics,* Fall, 129–44.

Rudin, Walter. 1973. *Functional analysis.* New York: McGraw-Hill.

Russell, Bertrand. 1912. *The problems of philosophy.* Oxford: Oxford University Press.

Schwartz, L. 1950. *Theorie des distributions.* Paris: Hermann.

Schwartz, Stephen P. 1977. (Ed.), *Naming, necessity, and natural kinds.* Ithaca, NY: Cornell University Press.

Shapiro, Stewart. 1983. Conservativeness and incompleteness. *Journal of Philosophy, 80,* 521–31.

1985. Second-order languages and mathematical practice. *Journal of Symbolic Logic, 50,* 714–42.

1989. Structure and ontology. *Philosophical Topics,* 145–71.

1990. Second-order logic, foundations, and rules. *Journal of Philosophy, 87,* 234–61.

Shoenfield, Joseph R. 1967. *Mathematical logic.* Reading, MA: Addison-Wesley.

Smith, D. E. 1958. *History of Mathematics,* Vol. 1. New York: Dover.

Soames, Scott. 1984. What is a theory of truth? *Journal of Philosophy, 81,* 411–29.

Spiegel, Murray R. 1964. *Theory and problems of complex variables.* Schaum's Outline Series. New York: McGraw-Hill.

Stroud, Barry. 1981. The significance of naturalized epistemology. In Peter A. French, Theodore E. Uehling, Jr., and Howard K. Wettstein (Eds.), *Midwest studies in philosophy* (Vol. 6, pp. 455–71). Minneapolis: University of Minnesota Press.

Tait, W. W. 1986. Truth and proof: The platonism of mathematics. *Synthese, 69,* 341–70.

Tarski, Alfred. 1944. The semantic conception of truth. *Philosophy and Phenomenological Research, 4,* 341–75.

1983. The concept of truth in formalized languages. In John Corcoran (Ed.), *Logic, semantics, metamathematics,* 2nd ed. (pp. 152–278). Indianapolis, IN: Hackett.

Troelstra, A. S. 1969. *Principles of intuitionism.* Berlin: Springer-Verlag.

Tymoczko, Thomas. 1979. The philosophical significance of the four color problem. *Journal of Philosophy, 76,* 57–84. Reprinted in Jay L. Garfield (Ed.), *Foundations of cognitive science* (pp. 139–62). New York: Paragon House, 1990.

1990. Why I am not a Turing machine: Gödel's theorems and the philosophy of mind. In Jay L. Garfield (Ed.), *Foundations of cognitive science* (pp. 170–85). New York: Paragon House.

Van Benthem, Johan, and Kees Doets. 1983. Higher-order logic. In D. Gabbay and F. Guenthner (Eds.), *Handbook of philosophical logic* (Vol. I, pp. 275–330). Dordrecht: Reidel.

Van Dalen, Dirk. 1986. Intuitionistic logic. In D. Gabbay and F. Guenthner (Eds.), *Handbook of philosophical logic* (Vol. II, pp. 225–329). Dordrecht: Reidel.

van der Waerden, B. L. 1963. *Science awakening*, trans. by Arnold Dresden. New York: Wiley.

Wagner, Steven J. 1982. Arithmetical fiction. *Pacific Philosophical Quarterly, 63,* 225–69.

1987. The rationalist conception of logic. *Notre Dame Journal of Formal Logic, 28,* 3–35.

Wallace, John. 1979. Only in the context of a sentence do words have any meaning. In Peter A. French, Theodore E. Uehling, Jr., and Howard K. Wettstein (Eds.), *Midwest studies in Philosophy,* Vol. 2. Minneapolis: University of Minnesota Press.

Wilkie, A. J., and J. B. Paris. 1987. On the scheme of induction for bounded arithmetic formulas. *Annals of Pure and Applied Logic, 35,* 261–302.

Wittgenstein, Ludwig. 1961. *Tractatus logico-philosophicus,* trans. by D. F. Pears and B. F. McGuinness. London: Humanities Press.

1968. *Philosophical investigations,* trans. by G. E. M. Anscombe. New York: Macmillan.

Yosida, Kosaku. 1978. *Functional analysis.* Berlin: Springer-Verlag.

Index

Kuhn, Thomas S., 52n

Lakatos, Imre, 5n, 25n, 26n, 54, 128n, 153n, 160n, 208n
Langer, R. E., 50, 51
languages, formal, 3–4, 5, 26, 68, 83
languages, natural, 3–4, 213n
Leblanc, Hughes, 92n, 221n
Leeds, Stephen, 91, 92n
Leibniz, G. W., 23, 50
Lewis, C. I., 46
Lewis, David, 67n
linguistic realism, 4, 5–6, 7, 148, 210
logic: boundary with mathematics, 3; centrality of, see Quine, centrality of logic and mathematics to conceptual scheme; compactness of first-order, 15, 222n; completeness of first-order, 15, 191; first-order, 3, 183; normativity of, 117–18; quantum, 190n; topic neutrality of, 3, 97, 114, 210; reasons to sacrifice topic neutrality of, 80–1, 103–4; irrelevance of topic neutrality of to evidential centrality, 189–90, 192; see also substitutional quantification; truth, as a disquotation device; truth predicate, Tarskian
Löwenheim-Skolem theorem, 3, 9, 11n, 15, 66
Luce, Lila, 71
Luker, Mark, 3n, 170n

Maddy, Penelope, 5n, 58, 59, 211; criticism of views, 32, 46, 47, 63, 64n; on Quine, 60, 71, 72; on set theory, 24n, 125n
Malament, D., 94n
Martin, Robert L., 89n
Mates, Benson, 226
mathematical authority, conditions for, 166
mathematical discourse, meaningfulness of, 148
mathematical justification, definition of, 166
mathematical necessity, 4
mathematical nominalism, 6, 7, 47–8, 55, 61, 62, 63–4, 212–13
mathematical objects: knowledge of, 5; reference to, 5, 7
mathematical practices: of ancient Egyptians, 21–2, 28; of the Andamans, 21–2; of the Babylonians, 21–2, 28; of the Greeks, 28
mathematical structuralism, 11, 133, 146
mathematical truth, centrality of, see Quine, centrality of logic and mathematics to conceptual scheme

mathematical truth, normativity of, 155–6
mathematics, language of, 4, 28
Mendelson, Elliot, 174n
Mill, J. S., 5n, 206
models, nonisomorphic, 3, 12, 16
monotonic connection, definition of, 122
monotonic overlap, definition of, 121
Moore, Gregory H., 125n, 128, 129n
Morse, Philip M., see Condon, Edward U.
Myhill, John, 3n, 8n

Nagel, Ernest, 23n
naturalized semantics, 7
negative numbers, 30n
Nelson, Edward, 56n
Nevanlinna, Rolf, and V. Paatero, 125n
Newman, James, 24n
Newton, Isaac, 50, 109n, 128
nominalism, see mathematical nominalism
nonfoundational set theory, 101
noun phrases, 4, 5, 6, 211

Occam's razor, 100–3
ω-logic, 9, 10, 12
ontologically conservative strategies, 7–11
ontologically radical strategies, 7–8

PA, 19, 82, 87, 122, 123, 126n, 130n, 132, 133, 135, 146, 211; nonstandard models of, 9; second-order, 11; standard model of, 140–3, 174
Paatero, V., see Nevanlinna, Rolf
Parikh, Rohit, 130n
Paris, J. B., see Wilkie, A. J.
Parsons, Charles, 5n, 11n, 64n, 135n, 198; criticism of Quine, 71, 74, 154, 175, 176, 177, 192
Pascal, B., 128
pb-system, definition of, 82
Peano arithmetic, see PA
Pearce, David, and Veikko Rantala, 8n
π, approximation of, 21
platonism: epistemological, 213n; methodological, 213n; ontological, 213n; traditional, 4n, 211–12
Popper, Karl, 89n
posits, local and nonlocal, definition of, 119
posits, thin and thick, 65
posits, ultrathin, definition of, 74
possession of an a priori warrant by a community of mathematicians, definition of, 163